The
Mind

ALSO BY JOHN BROCKMAN

AS AUTHOR
By the Late John Brockman
37
Afterwords
The Third Culture: Beyond the Scientific Revolution
Digerati

AS EDITOR
About Bateson
Speculations
Doing Science
Ways of Knowing
Creativity
The Greatest Inventions of the Past 2,000 Years
The Next Fifty Years
The New Humanists
Curious Minds
What We Believe but Cannot Prove
My Einstein
Intelligent Thought
What Is Your Dangerous Idea?
What Are You Optimistic About?
What Have You Changed Your Mind About?
This Will Change Everything
Is the Internet Changing the Way You Think?
Culture

AS COEDITOR
How Things Are (with Katinka Matson)

The
Mind

Leading Scientists Explore the Brain,
Memory, Personality, and Happiness

Edited by John Brockman

HARPER  PERENNIAL

NEW YORK • LONDON • TORONTO • SYDNEY • NEW DELHI • AUCKLAND

HARPER ● PERENNIAL

HarperCollins books may be purchased for educational, business, or sales promotional use. For information please write: Special Markets Department, HarperCollins Publishers, 10 East 53rd Street, New York, NY 10022.

FIRST EDITION

Library of Congress Cataloging-in-Publication Data

The mind : leading scientists explore the brain, memory, personality, and happiness / edited by John Brockman.
    p. cm.
  ISBN 978-0-06-202584-5 (pbk.)
  1. Brain. 2. Consciousness. 3. Neurosciences. I. Brockman, John, 1941–
QP376.M58275   2011
612.8—dc22

2010052196

11 12 13 14 15   OV/RRD   10 9 8 7 6 5 4 3 2 1

Contents

Introduction

In summer 2009, during a talk at the Ideas Festival in Bristol, England, physicist Freeman Dyson articulated a vision for the future. Responding to the recent book *The Age of Wonder*, in which Richard Holmes describes how the first Romantic Age was centered on chemistry and poetry, Dyson pointed out that today a new "Age of Wonder" has arrived that is dominated by computational biology. Its leaders include genomics researcher Craig Venter, medical engineer Dean Kamen, computer scientists Larry Page and Sergey Brin, and software architect and mathematician Charles Simonyi. The nexus for this intellectual activity, he observed, is online at Edge.org.

Dyson envisions an age of biology where "a new generation of artists, writing genomes as fluently as Blake and Byron wrote verses, might create an abundance of new flowers and fruit and trees and birds to enrich the ecology of our planet. Most of these artists would be amateurs, but they would be in close touch with science, like the poets of the earlier Age of Wonder. The new Age of Wonder might bring together wealthy entrepreneurs like Venter and Kamen . . . and a worldwide community of gardeners and farmers and breeders, working together to make the planet beautiful as well as fertile, hospitable to hummingbirds as well as to humans."

Indeed, Dyson was present at the August 2007 *Edge* meeting "Life: What a Concept," where he and genomics researchers Craig Venter and George Church, biologist Robert Shapiro, exo-biologist and astronomer Dimitar Sasselov, and quantum physicist Seth Lloyd presented their new, and in more than a few cases, startling research, and/or ideas in the biological sciences. "The meeting," according to *Suedduetsche Zeitung*, the largest national German newspaper, "was one of those memorable events that people in years to come will see as a crucial moment in history. After all, it's where the dawning of the age of biology was officially announced."

So, what is Edge.org?

First, *Edge* is people.

As the late artist James Lee Byars and I once wrote: "To accomplish the extraordinary, you must seek extraordinary people." At the center of every *Edge* publication and event are remarkable people and remarkable minds. *Edge*, at its core, consists of the scientists, artists, philosophers, technologists, and entrepreneurs who are at the center of today's intellectual, technological, and scientific landscape.

Second, *Edge* is events. Through its special lectures, Master Classes, and annual dinners in California, London, Paris, and New York, *Edge* gathers together the "third-culture" scientific intellectuals and technology pioneers who are exploring the themes of the post-industrial age. In this regard, commenting about the 2008 *Edge* Master Class "A Short Course in Behavioral Economics," science historian George Dyson wrote:

> Retreating to the luxury of Sonoma to discuss economic theory in mid-2008 conveys images of fiddling while Rome burns. Do the architects of Microsoft, Amazon, Google, PayPal, and Facebook have anything to teach the behavioral economists—and anything to learn? So what? What's new? As it turns out, all kinds of things are new. Entirely new economic structures and pathways have come into existence in the past few years.

It was a remarkable gathering of outstanding minds. These are the people that are rewriting our global culture.

Third, Edge.org is a conversation.

Edge is different from the Algonquin Roundtable or Bloomsbury Group, but it offers the same quality of intellectual adventure. Closer resemblances are the early 17th-century Invisible College, a precursor to the Royal Society. Its members consisted of scientists such as Robert Boyle, John Wallis, and Robert Hooke. The Society's common theme was to acquire knowledge through experimental investigation. Another inspiration is the Lunar Society of Birmingham, an informal club of the leading cultural figures of the new industrial age—James Watt, Erasmus Darwin, Josiah Wedgwood, Joseph Priestley, and Benjamin Franklin.

The online Edge.org salon is a living document of millions of words that charts the *Edge* conversation over the past fifteen years wherever it goes. It is available, gratis, to the general public.

Edge.org was launched in 1996 as the online version of the Reality Club, an informal gathering of intellectuals that met from 1981–1996 in Chinese restaurants, artist lofts, the board rooms of Rockefeller University and the New York Academy of Sciences, and investment banking firms, ballrooms, museums, living rooms, and elsewhere. Though the venue is now in cyberspace, the spirit of the Reality Club lives on in the lively back-and-forth conversations on the hot-button ideas driving the discussion today.

In the words of the novelist Ian McEwan, Edge.org is "open-minded, free ranging, intellectually playful . . . an unadorned pleasure in curiosity, a collective expression of wonder at the living and inanimate world . . . an ongoing and thrilling colloquium."

In this, the first volume of *The Best of* Edge *Series*, we focus on ideas about "Mind." We are pleased to present eighteen pieces, original works from the online pages of Edge.org, which consist of edited interviews, commissioned essays, and transcribed talks, many of which are accompanied online with streaming video. While there's no doubt about the value of online presentations, the role of books, whether bound and printed or presented electronically, is still an invaluable way to present important ideas. Thus, we are pleased to be able to offer this series of books to the public.

For this first volume, cutting-edge theoretical psychologists, cognitive scientists, neuroscientists, neurobiologists, linguists, behavioral geneticists, and moral psychologists explore new ways of thinking about "Mind."

In "Organs of Computation" (1997), Harvard psychologist Steven Pinker argues that "most of the assumptions about the mind that underlie current discussions are many decades out-of-date." He presents his idea that the basic understanding that the human mind is a remarkably complex processor of information, an "organ of extreme perfection and complication," to use Darwin's phrase, has not made it into the mainstream of intellectual life.

In "Philosophy in the Flesh" (1999), Berkeley cognitive scientist George Lakoff makes the point that "we are neural beings. Our brains take their input from the rest of our bodies. What our bodies are like and how they function in the world thus structures the very concepts we can use to think. We cannot think just anything—only what our embodied brains permit."

New York University neuroscientist Joseph Ledoux, in "Parallel Memories" (1997), argues for putting "emotion back into the brain and integrate it with cognitive systems. We shouldn't study emotion or cognition in isolation, but should study both as aspects of the mind in its brain."

Our minds evolved not as survival machines, but as courtship machines, says University of New Mexico psychologist Geoffrey Miller in "Sexual Selection and the Mind" (1998). He makes the point that "evolution is driven not just by natural selection for survival, but by an equally important process that Darwin called sexual selection through mate choice." He proposes that the human mind's most impressive, baffling abilities are courtship tools, evolved to attract and entertain sexual partners. By switching from a survival-centered view of evolution to a courtship-centered view, he attempts to show how we can understand the mysteries of mind.

Open University Emeritus Professor and neurobiologist Steven Rose is obsessed with the relationship between mind and brain. In "Rescuing Memory" (1999), he outlines his approach to understanding this relationship which has been to look for ways in which we can locate changes in behavior, thought, or action, which can be mapped in some way onto changes in physiology and biochemistry, and changes in structure in the brain, that is in processes that you can study biologically. For most of his life, the search has been focused on how we should understand learning and memory.

"During the last two decades," says evolutionary theorist Frank Sulloway in "How Is Personality Formed?" (1998), "I have experienced a major shift in my career interests. I started out as a historian of science and was primarily interested in historical questions about people's intellectual lives. In trying to understand the sources of creative achievement in science, I gradually became interested in problems of human development and especially in how Darwinian theory can help us to understand the development of personality. I now consider myself a psychologist, in addition to being an historian."

University of California at San Diego neuroscientist V. S. Ramachandran's widely cited essay "Mirror Neurons and Imitation Learning as the Driving Force Behind 'the Great Leap Forward' in Human Evolution" (2000) concerns "the discovery of mirror neurons in the frontal lobes of monkeys, and their potential relevance to human brain evolution—which I speculate on in this essay—is the single most important 'unreported' (or at least, unpublicized) story of the decade. I predict that mirror neurons will do for psychology what DNA did for biology: they will provide a unifying framework and help explain a host of mental abilities that have hitherto remained mysterious and inaccessible to experiments."

Theoretical psychologist Nicholas Humphrey, Emeritus Professor, London School of Economics, writes in his essay "A Self Worth Having" (2003) that "what I'm now thinking—though it certainly needs further work—is basically that the point of there being a phenomenally rich subjective present is that it provides a new domain for selfhood. Gottlob Frege, the great logician of the early 20th century, made the obvious but crucial observation that a first-person subject has to be the subject of something. In which case we can ask, what kind of something is up to doing the job? What kind of thing is of sufficient metaphysical weight to supply the experiential substrate of a self—or, at any rate, a self worth having? And the answer I'd now suggest is: nothing less than phenomenal experience—phenomenal experience with its intrinsic depth and richness, with its qualities of seeming to be more than any physical thing could be."

Stanford psychologist Philip Zimbardo, in "You Can't Be a Sweet Cucumber in a Vinegar Barrel" (2005), argues that "when you put that set of horrendous work conditions and external factors together, it creates an evil barrel. You could put virtually anybody in it and you're going to get this kind of evil behavior. The Pentagon and the military say that the Abu Ghraib scandal is the result of a few bad apples in an otherwise good barrel. That's the dispositional analysis. The social psychologist in me, and the consensus among many of my colleagues in experimental social psychology, says that's the wrong analysis. It's not the bad apples, it's the bad barrels that corrupt good people. Understanding the abuses at this Iraqi prison starts with an analysis of both the situational and systematic forces operating on those soldiers working the night shift in that 'little shop of horrors.'"

In V. S. Ramachandran's second essay "The Neurology of Self-Awareness" (2007), he writes, "What is the self? How does the activity of neurons give rise to the sense of being a conscious human being? Even this most ancient of philosophical problems, I believe, will yield to the methods of empirical science. It now seems increasingly likely that the self is not a holistic property of the entire brain; it arises from the activity of specific sets of interlinked brain circuits. But we need to know which circuits are critically involved and what their functions might be. It is the 'turning inward' aspect of the self—its recursiveness—that gives it its peculiar paradoxical quality."

"Eudaemonia: The Good Life" (2004) is University of Pennsylvania psychologist Martin Seligman's term for a "third form of happiness." This is meaning, which is "again knowing what your highest strengths are and deploying those in the service of something you believe is larger than you are. There's no shortcut to that. That's what life is about. There will likely be a pharmacology of pleasure, and there may be a pharmacology of positive emotion generally, but it's unlikely there'll be an interesting pharmacology of flow. And it's impossible that there'll be a pharmacology of meaning."

Collège de France experimental cognitive psychologist Stanislas Dehaene's "What Are Numbers, Really? A Cerebral Basis for Number Sense" (1997) presents research that number is very much like color. "Because we live in a world full of discrete and movable objects," he writes, "it is very useful for us to be able to extract number. This can help us to track predators or to select the best foraging grounds, to mention only very obvious examples. This is why evolution has endowed our brains and those of many animal species with simple numerical mechanisms. In animals, these mechanisms are very limited, as we shall see below: they are approximate, their representation becomes coarser for increasingly large numbers, and they involve only the simplest arithmetic operations (addition and subtraction). We, humans, have also had the remarkable good fortune to develop abilities for language and for symbolic notation. This has enabled us to develop exact mental representations for large numbers, as well as algorithms for precise calculations."

In "The Assortative Mating Theory" (2005) Cambridge psychologist Simon Baron-Cohen says that his thesis "with regard to sex differences is quite moderate, in that I do not discount environmental factors; I'm just saying, don't forget about biology. To me that sounds very moderate. But for some people in the field of gender studies, even that is too extreme. They want it to be all environment and no biology. You can understand that politically that was an important position in the 1960s, in an effort to try to change society. But is it a true description, scientifically, of what goes on? It's time to distinguish politics and science, and just look at the evidence."

Stanford biologist Robert Sapolsky, in "Toxo: The Parasite that Is Manipulating Human Behavior" (2009), notes that "the parasite my lab is beginning to focus on is one in the world of mammals, where parasites are changing mammalian behavior. It's got to do with this parasite, this protozoan called Toxoplasma. If you're ever pregnant, if you're ever around anyone who's pregnant, you know you immediately get skittish about cat feces, cat bedding, cat everything, because it could carry Toxo. And you do not want to get Toxoplasma into a fetal nervous system. It's a disaster."

"We've known for a long time that human children are the best learning machines in the universe, says Berkeley psychologist Alison Gopnik in "Amazing Babies" (2009), "but it has always been like the mystery of the humming birds. We know that they fly, but we don't know how they can possibly do it. We could say that babies learn, but we didn't know how."

Stanislas Dehaene presents the results of his recent research in "Signatures of Consciousness" (2009). Over the last twelve years, "my research team has been using all the brain research tools at its disposal, from functional MRI to electro- and magneto-encephalography and even electrodes inserted deep in the human brain, to shed light on the brain mechanisms of consciousness. I am now happy to report that we have acquired a good working hypothesis. In experiment after experiment, we have seen the same signatures of consciousness: physiological markers that all, simultaneously, show a massive change when a person reports becoming aware of a piece of information (say a word, a digit, or a sound)."

In "How Can Educated People Continue to Be Radical Environmentalists?" (1998), the late psychologist and behavioral geneticist David Lykken, formerly Emeritus Professor at the University of Minnesota, writes that "were it not for ideological prejudice any rational person looking at the evidence would agree that human aptitudes, personality traits, many interests and personal idiosyncrasies, even some social attitudes, owe from 30 to 70 percent of their variation across people to the genetic differences between people. The ideological barrier seems to involve the conviction that accepting these facts means accepting biological determinism, Social Darwinism, racism, and other evils." He draws on his famous study of 4,000 twins for comparisons between genetic and environmental effects on human psychology. "A better formula than Nature versus Nurture would be Nature *via* Nurture," he claims in support of his argument that the "genetic influences are strong and most of us develop along a path determined mainly by our personal genetic steersmen."

University of Virginia psychologist Jonathan Haidt explains in "Moral Psychology and the Misunderstanding of Religion" (2007) that "it might seem obvious to you that contractual societies are good, modern, creative, and free, whereas beehive societies reek of feudalism, fascism, and patriarchy. And, as a secular liberal I agree that contractual societies such as those of Western Europe offer the best hope for living peacefully together in our increasingly diverse modern nations (although it remains to be seen if Europe can solve its current diversity problems). I just want to make one point, however, that should give contractualists pause: surveys have long shown that religious believers in the United States are happier, healthier, longer-lived, and more generous to charity and to each other than are secular people."

John Brockman
Editor and Publisher
Edge.org

The
Mind

1

Organs of Computation

Steven Pinker

Johnstone Family Professor, Department of Psychology, Harvard University; Author, The Language Instinct, The Blank Slate, *and* The Stuff of Thought

EDGE: How does one even begin to explain something as complicated as the human mind?

STEVEN PINKER: I think the key to understanding the mind is to try to "reverse engineer" it—to figure out what natural selection designed it to accomplish in the environment in which we evolved. In my book *How the Mind Works*, I present the mind as a system of "organs of computation" that allowed our ancestors to understand and outsmart objects, animals, plants, and each other.

EDGE: How is that approach different from what intellectuals currently believe?

PINKER: Most of the assumptions about the mind that underlie current discussions are many decades out of date. Take the hydraulic model of Freud, in which psychic pressure builds up in the mind and can burst out unless it's channeled into appropriate pathways. That's just false. The mind doesn't work by fluid under pressure or by flows of energy; it works by information. Or, look at the commentaries on human affairs by pundits and social critics. They say we're "conditioned" to do this, or "brainwashed" to do that, or "socialized" to believe such and such. Where do these ideas come from? From the behaviorism of the 1920s, from bad cold war movies from the 1950s, from folklore about the effects of family upbringing that behavior genetics has shown to be false. The basic understanding that the human mind is a remarkably complex processor of information, an "organ of extreme perfection and complication," to use Darwin's phrase, has not made it into the mainstream of intellectual life.

EDGE: What makes you say that the mind is such a complex system?

PINKER: What should impress us about the mind is not its rare extraordinary feats, like the accomplishments of Mozart or Shakespeare or Einstein, but the everyday feats we take for granted. Seeing in color. Recognizing your mother's face. Lifting a milk carton and gripping it just tight enough that it doesn't drop but not so tight that you crush it, while rocking it back and forth to gauge how much milk is in the bottom just from the tugs on your fingertips. Reasoning about the world—what will and won't happen when you open the refrigerator door. All of these things sound mundane and boring, but they shouldn't be. We can't, for example, program a robot to do any of them! I would pay a lot for a robot that would put away the dishes or run simple errands, but I can't, because all of the little problems that you'd need to solve to build a robot to do that, like recognizing objects, reasoning about the world, and controlling hands and feet, are unsolved engineering problems. They're much harder than putting a man on the moon or sequencing the human genome. But a four-year-old solves them every time she runs across the room to carry out an instruction from her mother.

I see the mind as an exquisitely engineered device—not literally engineered, of course, but designed by the mimic of engineering that we see in nature, natural selection. That's what "engineered" animals' bodies to accomplish improbable feats, like flying and swimming and running, and it is surely what "engineered" the mind to accomplish its improbable feats.

EDGE: What does that approach actually buy you in studying how the mind works?

PINKER: It tells you what research in psychology should be: a kind of reverse engineering. When you rummage through an antique store and come across a contraption built of many finely meshing parts, you assume that it was put together for a purpose, and that if you only understood that purpose, you'd have insight as to why it has the parts arranged the way they are. That's true for the mind as well, though it wasn't designed by a designer but by natural selection. With that insight you can look at the quirks of the mind and ask how they might have made sense as solutions to some problem our ancestors faced in negotiating the world. That can give you an insight into what the different parts of the mind are doing.

Even the seemingly irrational parts of the mind, like strong passions—jealousy, revenge, infatuation, pride—might very well be good solutions to problems our ancestors faced in dealing with one another. For example, why do people do crazy things like chase down an ex-lover and kill the lover? How could you win someone back by killing them? It seems like a bug in our mental software. But several economists have proposed an alternative. If our mind is put together so that under some circumstances we are compelled to carry out a threat regardless of the costs to us, the threat is made credible. When a person threatens a lover, explicitly or implicitly, by communicating, "If you ever leave me I'll chase you down," the lover could call his bluff if she didn't have signs that he was crazy enough to carry it out even though it was pointless. And so the problem of building a credible deterrent into creatures that interact with one another leads to irrational behavior as a rational solution. "Rational," that is, with respect to the "goal" of our genes to maximize the number of copies of themselves. It isn't "rational," of course, with respect to the goal of whole humans and societies to maximize happiness and fairness.

Another example is the strange notion of happiness. What is the psychological state called "happiness" for? It can't be that natural selection designed us to feel good all the time out of sheer goodwill. Presumably our brain circuits for happiness motivate us to accomplish things that enhance biological fitness. With that simple insight one can make some sense of some of the puzzles of happiness that wise men and women have noted for thousands of years. For example, directly pursuing happiness is often a recipe for unhappiness, because our sense of happiness is always calibrated with respect to other people. There is a Yiddish expression: When does a hunchback rejoice? When he sees one with a bigger hump.

Perhaps we can make sense of this by putting ourselves in the shoes of the fictitious engineer behind natural selection. What should the circuit for happiness be doing? Presumably it would be assessing how well you're doing in your current struggle in life—whether you should change your life and try to achieve something different, or whether you should be content with what you've achieved so far, for example, when you are well-fed, comfortable, with a mate, in a situation likely to result in children, and so on. But how could a brain be designed in advance to assess that? There's no absolute standard for well-being. A Paleolithic hunter-gatherer should not have fretted that he had no running shoes or central heating or penicillin. How can a brain know whether there is something worth striving for? Well, it can look around and see how well-off other people are. If they can achieve something, maybe so can you. Other people anchor your well-being scale and tell you what you can reasonably hope to achieve.

Unfortunately, it gives rise to a feature of happiness that makes many people unhappy—namely, you're happy when you do a bit better than everyone around you and you're unhappy when you're doing worse. If you look in your paycheck envelope and you discover you got a 5 percent raise you'd be thrilled, but if you discover that all your coworkers got a 10 percent raise you'd be devastated.

Another paradox of happiness is that losses are felt more keenly than gains. As Jimmy Connors said, "I hate to lose more than I like to win." You are just a little happy if your salary goes up, but you're really miserable if your salary goes down by the same amount. That

Steven Pinker

too might be a feature of the mechanism designed to attain the attainable and no more. When we backslide, we keenly feel it because what we once had is a good estimate of what we can attain. But when we improve we have no grounds for knowing that we are as well-off as we can hope to be. The evolutionary psychologist Donald Campbell called it "the happiness treadmill." No matter how much you gain in fame, wealth, and so on, you end up at the same level of happiness you began with—though to go down a level is awful. Perhaps it's because natural selection has programmed our reach to exceed our grasp, but by just a little bit.

EDGE: How do you differ from other people who have written about the mind, like Dan Dennett, John Searle, Noam Chomsky, Gerald Edelman, or Francis Crick?

PINKER: For starters, I place myself among those who think that you can't understand the mind only by looking directly at the brain. Neurons, neurotransmitters, and other hardware features are widely conserved across the animal kingdom, but species have very different cognitive and emotional lives. The difference comes from the ways in which hundreds of millions of neurons are wired together to process information. I see the brain as a kind of computer—not like any commercial computer made of silicon, obviously, but as a device that achieves intelligence for some of the same reasons that a computer achieves intelligence—namely, processing of information. That places me with Dennett and Chomsky (though the three of us disagree about much else), and in disagreement with people like Searle, who denies that the brain can be understood as an information processor and insists it can only be understood in terms of physiology. Edelman and Crick would not state their views in terms as extreme as Searle's but they, too, are not entirely sympathetic to the computational theory of mind.

Like Dennett and Searle, but unlike Chomsky, I believe that natural selection is the key to explaining the structure of the mind—that reverse engineering in the light of natural selection is the key to answering why our thoughts and feelings are structured as they are.

I also believe that the mind is not made of Spam—it has a complex, heterogeneous structure. It is composed of mental organs that are specialized to do different things, like seeing, controlling hands and feet, reasoning, language, social interaction, and social emotions. Just as the body is divided into physical organs, the mind is divided into mental organs. That puts me in agreement with Chomsky and against many neural network modelers, who hope that a single kind of neural network, if suitably trained, can accomplish every mental feat that we do. For similar reasons I disagree with the dominant position in modern intellectual life—that our thoughts are socially constructed by how we were socialized as children, by media images, by role models, and by conditioning.

EDGE: But haven't there been objections to the computer metaphor of the mind?

PINKER: Some critics think it is an example of our mindless incorporating the latest technology into our theories. The objection goes: When telephone switchboards first came into existence, people thought the mind was a switchboard; before that, when fancy water-powered mechanical toys were the rage, people said the mind was a hydraulic machine, and so on. Of course there's a danger in taking metaphors too literally, but when you're careful, mechanical metaphors really do increase our understanding. The heart and blood vessels really can be better understood by thinking about pumps and pipes, and the switchboard metaphor offers a clearer understanding of the nerves and spinal cord than the models that came before it.

And I think the theory of computation, and in some cases real computers, do offer principles that are essential to understanding how the mind works. The idea is not that the mind is like a commercial computer; it's that minds and computers work by some of the same principles. When engineers first came to understand flight

Steven Pinker

as they designed airplanes, it provided insight as to how birds fly, because principles of aerodynamics, like the shape of an airfoil or the interplay of lift and drag, are applicable both to planes and to birds. That doesn't mean that the airplane is a good model of the birds. Birds don't have propellers and headphone jacks and beverage service, for example. But by understanding the laws that allow any device to fly, one can understand how natural devices fly. The human mind is unlike a computer in countless ways, but the trick behind computation is the trick behind thought—representing states of the world; that is, recording information, and manipulating the information according to rules that mimic relations of truth and statistical probability that hold in the world.

EDGE: Haven't there also been political objections to the biological approach you are taking?

PINKER: Many people lump together the idea that the mind has a complex innate structure with the idea that differences between people have to be innate. But the ideas are completely different. Every normal person on the planet could be innately equipped with an enormous catalog of mental machinery, and all the differences between people—what makes John different from Bill—could come from differences in experience, of upbringing, or of random things that happened to them when they were growing up. To believe that there's a rich innate structure common to every member of the species is different from saying the differences between people, or differences between groups, come from differences in innate structure. Here's an example: Look at number of legs—it's an innate property of the human species that we have two legs as opposed to six like insects, or eight like spiders, or four like cats—so having two legs is innate. But if you now look at why some people have one leg, and some people have no legs, it's completely due to the environment—they lost a leg in an accident, or from a disease. So the two questions have to be distinguished. And what's true of legs is also true of the mind.

EDGE: As you know, I have been increasingly interested in the growing presence of the Internet and its effects on intellectual life. Do you think that what we know about the mind has any implications for how quickly computer technology will change our world?

PINKER: Computer technology will never change the world as long as it ignores how the mind works. Why did people instantly start to use fax machines, and continue to use them even though electronic mail makes much more sense? There are millions of people who print out text from their computer onto a piece of paper, feed the paper into a fax machine, forcing the guy at the other end to take the paper out, read it, and crumple it up—or worse, scan it into his computer so that it becomes a file of bytes all over again. This is utterly ridiculous from a technological point of view, but people do it. They do it because the mind evolved to deal with physical objects, and it still likes to conceptualize entities that are owned and transferred among people as physical objects that you can lift and store in a box. Until computer systems, e-mail, video cameras, VCRs, and so on are designed to take advantage of the way the mind conceptualizes reality—namely, as physical objects existing at a location and impinged upon by forces—people are going to be baffled by their machines, and the promise of the computer revolution will not be fulfilled.

Part of the problem may be that our best technology comes from Japan and the manuals were written in Japanese and then translated, but I have a hunch that in Japan they have as much trouble programming the VCR as we do here. It's not just the instructions, but the design of the machines themselves, that's the problem. The machines were designed by engineers who aren't used to thinking about how the human mind works. They're used to designing machinery that is elegant by their own standards, and they don't think about how the user is going to conceptualize the machine as another object in the world and deal with it as we've been dealing with objects for hundreds of thousands of years.

Steven Pinker

EDGE: Let me turn the question around. What is the significance of the Internet and today's communications revolution for the evolution of the mind?

PINKER: Probably not much. You've got to distinguish two senses of the word "evolution." The sense used by me, Dawkins, Gould, and other evolutionary biologists refers to the changes in our biological makeup that led us to be the kind of organism we are today. The sense used by most other people refers to continuous improvement or progress. A popular idea is that our biological evolution took us to a certain stage, and our cultural evolution is going to take over—where evolution in both cases is defined as "progress." I would like us to move away from that idea, because the processes that selected the genes that built our brains are different from the processes that propelled the rise and fall of empires and the march of technology.

In terms of strict biological evolution, it's impossible to know where, if anywhere, our species is going. Natural selection generally takes hundreds of thousands of years to do anything interesting, and we don't know what our situation will be like in 10,000 or even 1,000 years. Also, selection adapts an organism to a niche, usually a local environment, and the human species moves all over the place and lurches from lifestyle to lifestyle with dizzying speed on the evolutionary timetable. Revolutions in human life like the agricultural, industrial, and information revolutions occur so quickly that no one can predict what change they will have on our makeup, or even whether there will be a change.

The Internet does create a kind of supra-human intelligence, in which everyone on the planet can exchange information rapidly, a bit like the way different parts of a single brain can exchange information. This is not a new process; it's been happening since we evolved language. Even nonindustrial hunter-gatherer tribes pool information by the use of language. That has given them remarkable local technologies—ways of trapping animals, using poisons, chemically treating plant foods to remove the bitter toxins, and so on. That is also a collective intelligence that comes from accumulating discoveries over generations, and pooling them among a group of people living at one time. Everything that's happened since, such as writing, the printing press, and now the Internet, are ways of magnifying something that our species already knew how to do, which is to pool expertise by communication. Language was the real innovation in our biological evolution; everything since has just made our words travel farther or last longer.

Steven Pinker

2

Philosophy in the Flesh

George Lakoff

Cognitive Scientist and Linguist, Richard and Rhoda Goldman Distinguished Professor of Cognitive Science and Linguistics, University of California, Berkeley; Author, Don't Think of an Elephant *and* The Political Mind

EDGE: What is a body?

GEORGE LAKOFF: That's an interesting question. Pierre Bourdieu has pointed out that our bodies and what we do with them differ significantly from culture to culture. Frenchmen do not walk like Americans do. Women's bodies are different than men's bodies. The Chinese body is not like the Polish body. And our understanding of what the body is has changed drastically over time, as postmodernists have often observed.

But nonetheless, our bodies do share a lot. We have two eyes, two ears, two arms, two legs, blood that circulates, lungs used to breathe, skin, internal organs, and on and on. The common conventionalized aspects of our conceptual systems tend to be structured by what our bodies have in common, which is a lot.

EDGE: But we go from being a machine to an information system, and eventually those orifices may not be part of the conversation.

LAKOFF: When you start to study the brain and body scientifically, you inevitably wind up using metaphors. Metaphors for the mind, as you say, have evolved over time—from machines to switchboards to computers. There's no avoiding metaphor in science. In our lab, we use the neural circuitry metaphor ubiquitous throughout neuroscience. If you're studying neural computation, that metaphor is necessary. In the day-to-day research on the details of neural computation, the biological brain moves into the background while the neural circuitry introduced by the metaphor is what one works with. But no matter how ubiquitous a metaphor may be, it is important to keep track of what it hides and what it introduces. If you don't, the body does disappear. We're careful about our metaphors, as most scientists should be.

EDGE: There were no information-processing metaphors thirty-five to forty years ago—and so is the body real, or is it invented?

LAKOFF: There's a difference between the body and our conceptualization of it. The body is the same as it was thirty-five years ago; the conception of the body is very different. We have metaphors for the body we didn't have then, with relatively advanced science built on those metaphors. In this respect, the contemporary body and brain, conceptualized in terms of neural circuitry and other information-processing metaphors, were "invented." Such inventions are crucial to science. Our emerging understanding of the embodiment of mind would not be possible without them.

EDGE: How does this approach depart from your early work?

LAKOFF: My really early work was done between 1963 and 1975, when I was pursuing the theory of generative semantics. During that period, I was attempting to unify Chomsky's transformational grammar with formal logic. I had helped work out a lot of the early details of Chomsky's theory of grammar. Noam claimed then—and still does, so far as I can tell—that syntax is independent of meaning, context, background knowledge, memory, cognitive processing, communicative intent, and every aspect of the body.

In working through the details of his early theory, I found quite a few cases where semantics, context, and other such factors entered into rules governing the syntactic occurrences of phrases and morphemes. I came up with the beginnings of an alternative theory in 1963 and, along with wonderful collaborators like Haj Ross and Jim

George Lakoff

McCawley, developed it through the sixties. Back in 1963, semantics meant logic—deductive logic and model theory—and our group developed a theory of generative semantics that united formal logic and transformational grammar. In that theory, semantics (in the form of logic) was taken as prior to syntax on the basis of evidence that semantic and pragmatic considerations entered into generalizations governing syntactic structure. Chomsky has since adopted many of our innovations, though he fought them viciously in the 1960s and 1970s.

In 1975, I became acquainted with certain basic results from the various cognitive sciences pointing toward an embodied theory of mind—the neurophysiology of color vision, prototypes and basic-level categories, Talmy's work on spatial relations concepts, and Fillmore's frame semantics. These results convinced me that the entire thrust of research in generative linguistics and formal logic was hopeless. I set about, along with Len Talmy, Ron Langacker, and Gilles Fauconnier, to form a new linguistics—one compatible with research in cognitive science and neuroscience. It is called Cognitive Linguistics, and it's a thriving scientific enterprise. In 1978, I discovered that metaphor was not a minor kind of trope used in poetry, but rather a fundamental mechanism of mind. In 1979, Mark Johnson visited in the Berkeley Philosophy Department and we began working out the details and their implications for philosophy. We've been collaborating for twenty years. Mark is now Chair of Philosophy at Oregon.

EDGE: Distinguish cognitive science from philosophy.

LAKOFF: That is a deep and important question, and central to the enterprise of *Philosophy in the Flesh*. The reason that the question doesn't have a simple answer is that there are two forms of cognitive science, one fashioned on the assumptions of Anglo-American philosophy and one (so far as we can tell) independent of specific philosophical assumptions that determine the results of the inquiry.

Early cognitive science, what we call "first-generation" cognitive science (or "disembodied cognitive science"), was designed to fit a formalist version of Anglo-American philosophy. That is, it had philosophical assumptions that determined important parts of the content of the scientific "results." Back in the late 1950s, Hilary Putnam (a noted and very gifted philosopher) formulated a philosophical position called "functionalism." (Incidentally, he has since renounced that position.) It was an a priori philosophical position, not based on any evidence whatever.

The proposal was this: The mind can be studied in terms of its cognitive functions—that is, in terms of the operations it performs—independently of the brain and body. The operations performed by the mind can be adequately modeled by the manipulation of meaningless formal symbols, as in a computer program. This philosophical program fit paradigms that existed at the time in a number of disciplines.

In formal philosophy: the idea that reason could be adequately characterized using symbolic logic, which utilizes the manipulation of meaningless formal symbols.

In generative linguistics: the idea that the grammar of a language can be adequately characterized in terms of rules that manipulate meaningless formal symbols.

In artificial intelligence: the idea that intelligence in general consists in computer programs that manipulate meaningless formal symbols.

In information-processing psychology: the idea that the mind is an information-processing device, where information processing is taken as the manipulation of meaningless formal symbols, as in a computer program.

All of these fields had developed out of formal philosophy. These four fields converged in the 1970s to form first-generation cognitive science. It had a view of mind as the disembodied manipulation of meaningless formal symbols.

EDGE: How does this fit into empirical science?

LAKOFF: This view was not empirically based, having arisen from an a priori philosophy. Nonetheless it got the field started. What was good about it was that it was precise. What was disastrous about it was that it had a hidden philosophical worldview that masqueraded as a scientific result. And if you accepted that philosophical position, all results inconsistent with that philosophy could only be seen as nonsense. To researchers trained in that tradition, cognitive science was the study of mind within that a priori philosophical position. The first generation of cognitive scientists was trained to think that way, and many textbooks still portray cognitive science in that way. Thus, first-generation cognitive science is not distinct from philosophy; it comes with an a priori philosophical worldview that places substantive constraints on what a "mind" can be. Here are some of those constraints:

Concepts must be literal. If reasoning is to be characterized in terms of traditional formal logic, there can be no such thing as a metaphorical concept and no such thing as metaphorical thought.

Concepts and reasoning with concepts must be distinct from mental imagery, since imagery uses the mechanisms of vision and cannot be characterized as being the manipulation of meaningless formal symbols.

Concepts and reasoning must be independent of the sensory-motor system, since the sensory-motor system, being embodied, cannot be a form of disembodied abstract symbol manipulation.

Language, too—if it was to fit the symbol-manipulation paradigm—had to be literal, independent of imagery, and independent of the sensory-motor system.

From this perspective, the brain could only be a means to implement abstract "mind"—wetware on which the "programs of the mind" happened to be implementable. Mind on this view does not arise from and is not shaped by the brain. Mind is a disembodied abstraction that our brains happen to be able to implement. These were not empirical results, but rather followed from philosophical assumptions.

In the mid–1970s, cognitive science was finally given a name and outfitted with a society and a journal. The people who formed the field accepted the symbol-manipulation paradigm. I was originally one of them (on the basis of my early work on generative semantics) and gave one of the invited inaugural lectures at the first meeting of the Cognitive Science Society. But just around the time that the field officially was recognized and organized around the symbol-manipulation paradigm, empirical results started coming in calling the paradigm itself into question.

This startling collection of results pointed toward the idea that mind was not disembodied—not characterizable in terms of the manipulation of meaningless symbols independent of the brain and body; that is, independent of the sensory-motor system and our functioning in the world. Mind instead is embodied, not in the trivial sense of being implementable in a brain, but in the crucial sense that conceptual structure and the mechanisms of reason arise ultimately and are shaped by the sensory-motor system of the brain and body.

EDGE: Can you prove it?

LAKOFF: There is a huge body of work supporting this view. Here are some of the basic results that have interested me the most: The structure of the system of color categories is shaped by the neurophysiology of color vision, by our color cones and neural circuitry for color. Colors and color categories are not "out there" in the world but are interactional, a nontrivial product of wavelength reflectances of objects and lighting conditions on the one hand, and our color cones and neural circuitry on the other. Color concepts and color-based inferences are thus structured by our bodies and brains.

Basic-level categories are structured in terms of gestalt perception, mental imagery, and motor schemas. In this way the body and the

George Lakoff

sensory-motor system of the brain enters centrally into our conceptual systems.

Spatial relations concepts in languages around the world (for example, in, through, around in English, *sini* in Mixtec, *mux* in Cora, and so on) are composed of the same primitive "image schemas"— that is, schematic mental images. These, in turn, appear to arise from the structure of visual and motor systems. This forms the basis of an explanation of how we can fit language and reasoning to vision and movement.

Aspectual concepts (which characterize the structure of events) appear to arise from neural structures for motor control.

Categories make use of prototypes of many sorts to reason about the categories as a whole. Those prototypes are characterized partly in terms of sensory-motor information.

The conceptual and inferential system for reasoning about bodily movements can be performed by neural models that can model both motor control and inference. Abstract concepts are largely metaphorical, based on metaphors that make use of our sensory-motor capacities to perform abstract inferences. Thus, abstract reason, on a large scale, appears to arise from the body.

These are the results most striking to me. They require us to recognize the role of the body and brain in human reason and language. They thus run contrary to any notion of a disembodied mind. It was for such reasons that I abandoned my earlier work on generative semantics and started studying how mind and language are embodied. They are among the results that have led to a second generation of cognitive science, the cognitive science of the embodied mind.

EDGE: Let's get back to my question about the difference between cognitive science and philosophy.

LAKOFF: OK. Cognitive science is the empirical study of the mind, unfettered by a priori philosophical assumptions. First-generation cognitive science, which posed a disembodied mind, was carrying out a philosophical program. Second-generation cognitive science, which is working out the nature of the mind as it really is—embodied!—had to overcome the built-in philosophy of earlier cognitive science.

EDGE: Does "second-generation cognitive science" presuppose a philosophy?

LAKOFF: We have argued that it does not, that it simply presupposes commitments to take empirical research seriously, seek the widest generalizations, and look for convergent evidence from many sources. That is just what science is committed to. The results about the embodied mind did not begin from, and do not presuppose, any particular philosophical theory of mind. Indeed, it has required separating out the old philosophy from the science.

EDGE: Where does this leave philosophy?

LAKOFF: In a position to start over from an empirically responsible position. Young philosophers should be thrilled. Philosophy is anything but dead. It has to be rethought taking the empirical results about the embodied mind into account. Philosophy considers the deepest questions of human existence. It is time to rethink them and that is an exciting prospect.

EDGE: What about the academic wars between postmodern and analytic philosophy?

LAKOFF: The results suggest that both sides were insightful in some respects and mistaken in others. The postmodernists were right that some concepts can change over time and vary across cultures. But they were wrong in suggesting that all concepts are like that. Thousands are not. They arise around the world in culture after culture from our common embodiment.

Postmodernists were right in observing that there are many places where the folk theory of essences fails. But they were wrong in suggesting that such a failure undercuts our conceptual systems and makes them arbitrary. The analytic tradition insightfully characterized the theory of speech acts. Although formal logic does not work for all, or even most, of reason, there are places where something akin to formal logic (much revised) does characterize certain limited aspects of reason. But the analytic tradition was wrong in certain of its central theses: the correspondence theory of truth, the theory of literal meaning, and the disembodied nature of reason.

The academic world is now in a position to transcend both positions, each having contributed something important and each needing revision.

EDGE: Is there an East Coast and West Coast divide?

LAKOFF: Dan Dennett referred to the "East Pole" and "West Pole" back in the early to mid 1980s, as if the proponents of the disembodied mind were all on the East Coast and the proponents of the embodied mind were all on the West Coast. Research on the embodied mind did tend to start on the West Coast, but even then the geographical characterization was oversimplified. By now, both positions are represented on both coasts and throughout the country. Cambridge and Princeton in the past have tended largely toward the old disembodied mind position, at least in certain fields. But there are so many interesting thinkers on both coasts and spread across the country that I think that any geographical divisions that still exist won't last long.

When Dennett first made that distinction, the great revolutions in neuroscience and neural modeling were just starting. Cognitive linguistics was just coming into existence. *Metaphors We Live By* had barely come out and *Women, Fire, and Dangerous Things* had not yet been written. Nor had Edelman's *Bright Air, Brilliant Fire*, nor Damasio's *Descartes' Error*, nor Regier's *The Human Semantic Potential*, nor the various books by Pat and Paul Churchland. Over the past decade and a half, neuroscience and neural computation have changed the landscape of cognitive science and they will change it even more in the next decade or two. Those changes will inevitably move us further toward an appreciation of the embodiment of mind. You cannot think anything without using the neural system of your brain. The fine structure of neural connections in the brain, their connections to the rest of the body, and the nature of neural computation will keep being developed. The more we discover about the details, the more we will come to understand the detailed nature of how reason and the conceptual systems in which we reason are embodied.

The idea of disembodied reason was an a priori philosophical idea. It lasted 2,500 years. I can't imagine it lasting another 30 years in serious scientific circles.

EDGE: And what do we have to look forward to?

LAKOFF: Cognitive science and neuroscience are triggering a philosophical revolution. *Philosophy in the Flesh* is just part of the first wave. Over the next decade or two, the neural theory of language should develop sufficiently to replace the old view of language as meaningless disembodied symbol manipulation that one finds in the old Chomskyan tradition. But the biggest, and one of the most important, changes will come in our understanding of mathematics.

The precursor of that change was Stanislas Dehaene's *The Number Sense*, which reviewed the evidence from neuroscience, child development, and animal research indicating that we (and certain other animals) have evolved with a part of our brains dedicated to enumeration and simple arithmetic up to a small number of objects (around four). Rafael Núñez and I begin with those findings and ask how sophisticated arithmetic (with the laws of arithmetic) developed—that is, how could ordinary conceptual mechanisms for human thought have given rise to mathematics?

Our answer is that the ordinary embodied mind, with its image schemas, conceptual metaphors, and mental spaces, has the capacity to create the most sophisticated of mathematics via using everyday conceptual mechanisms. Dehaene stopped with simple arithmetic. We go on to show that set theory, symbolic logic, algebra, analytic geometry, trigonometry, calculus, and complex numbers can all be accounted for using those everyday conceptual mechanisms. Moreover, we show that conceptual metaphor is at the heart of the development of complex mathematics. This is not hard to see. Think of the number line. It is the result of a metaphor that Numbers Are Points on a Line. Numbers don't have to be thought of as points on a line. Arithmetic works perfectly well without being thought of in terms of geometry. But if you use that metaphor, much more interesting mathematics results. Or take the idea, in set-theoretical foundations for arithmetic, that Numbers Are Sets, with zero as the empty set, one as the set containing the empty set, and so on. That's a metaphor, too. Numbers don't have to be thought of as being sets. Arithmetic went on perfectly well for 2,000 years without numbers being conceptualized as sets. But if you use that metaphor, then interesting mathematics results. There is a third less well-known metaphor for numbers, that Numbers Are Values of Strategies in combinatorial game theory. So which is it? Are numbers points? Are they sets? Are numbers fundamentally just values of strategies in combinatorial games?

These metaphors for numbers are part of the mathematics, and you make a choice each time depending on the kind of mathematics you want to be doing. The moral is simple: Conceptual metaphor is central to conceptualization of number in mathematics of any complexity at all. It's a perfectly sensible idea. Conceptual metaphors are cross-domain mappings that preserve inferential structure. Mathematical metaphors are what provide the links across different branches of mathematics. One of our most interesting results concerns the conceptualization of infinity. There are many concepts that involve infinity: points at infinity in projective and inversive geometry, infinite sets, infinite unions, mathematical induction, transfinite numbers, infinite sequences, infinite decimals, infinite sums, limits, least upper bounds, and infinitesimals. Núñez and I have found that all of these concepts can be conceptualized as special cases of one simple Basic Metaphor of Infinity. The idea of "actual infinity"—of infinity not just as going on and on, but as a thing—is metaphorical, but the metaphor, as we show, turns out to be quite simple and exists outside of mathematics. What mathematicians have done is to provide elaborate, carefully devised special cases of this basic metaphorical idea.

What we conclude is that mathematics as we know it is a product of the human body and brain; it is not part of the objective structure of the universe—this or any other. What our results appear to disprove is what we call the Romance of Mathematics, the idea that mathematics exists independently of beings with bodies and brains and that mathematics structures the universe independently of any embodied beings to create the mathematics. This does not, of course, result in the idea that mathematics is an arbitrary product of culture, as some postmodern theorists would have it. It simply says that it is a stable product of our brains, our bodies, our experience in the world, and aspects of culture. The explanation of why mathematics "works so well" is simple: It is the result of tens of thousands of very smart people observing the world carefully and adapting or creating mathematics to fit their observations. It is also the result of a mathematical evolution: a lot of mathematics invented to fit the world turned out not to. The forms of mathematics that work in the world are the result of such an evolutionary process.

It is important to know that we create mathematics and to understand just what mechanisms of the embodied mind make mathematics possible. It gives us a more realistic appreciation of our role in the universe. We, with our physical bodies and brains, are the source of reason, the source of mathematics, the source of ideas. We are not mere vehicles for disembodied concepts, disembodied reason, and disembodied mathematics floating out there in the universe. That makes each embodied human being (the only kind) infinitely valuable—a source, not a vessel. It makes bodies infinitely valuable— the source of all concepts, reason, and mathematics.

For two millennia, we have been progressively devaluing human

George Lakoff

life by underestimating the value of human bodies. We can hope that the next millennium, in which the embodiment of mind will come to be fully appreciated, will be more humanistic.

EDGE: Where are you headed next?

LAKOFF: I've plunged myself as fully as possible into the research that Jerry Feldman and I have been doing for the past decade at the International Computer Science Institute on the Neural Theory of Language. That's where most of my technical research effort is going to go for quite a while.

Jerry developed the theory of structured connectionism (not PDP connectionism) beginning in the 1970s. Structured connectionism allows us to construct detailed computational neural models of conceptual and linguistic structures and of the learning of such structures.

Since 1988, we've been running a project that takes up a question that has absorbed both of us: From the perspective of neural computation, a human brain consists of a very large number of neurons connected up in specific ways with certain computational properties. How is it possible to get the details of human concepts, the forms of human reason, and the range of human languages out of a lot of neurons connected up as they are in our brains? How do you get thought and language out of neurons? That is the question we are trying to answer in our lab through the computational neural modeling of thought and language.

EDGE: How do you connect structures in the brain to ideas of space?

LAKOFF: Terry Regier has taken the first step to figuring that out in his book *The Human Semantic Potential*. He has hypothesized that certain types of brain structures—topographic maps of the visual field, orientation-sensitive cells, and so on—can compute the primitive spatial relations (called "image schemas") that linguists have discovered. The amazing thing to me is that not only do we actually have a reasonable idea of how certain types of neural structures can give rise to spatial relations concepts. Recent neural modeling research by Srini Narayanan has similarly given us an idea of how brain structures can compute aspectual concepts (which structure events), conceptual metaphors, mental spaces, blended spaces, and other basics of human conceptual systems. The next breakthrough, I think, will be a neural theory of grammar.

These are remarkable technical results. When you put them together with other results about the embodiment of mind coming from neuroscience, psychology, and cognitive linguistics, they tell us a great deal about things that are important in the everyday lives of ordinary people—things that philosophers have speculated about for over 2,500 years. Cognitive science has important things to tell us about our understanding of time, events, causation, and so on.

EDGE: Like what?

LAKOFF: When Mark Johnson and I looked over these results from the cognitive sciences in detail, we realized that there were three major results that were inconsistent with almost all of Western philosophy (except for Merleau-Ponty and Dewey), namely:

The mind is inherently embodied.

Most thought is unconscious.

Abstract concepts are largely metaphorical.

This realization led us to ask the following question in *Philosophy in the Flesh*: What would happen if we started with the new results about the mind and reconstructed philosophy from there? What would philosophy look like?

It turns out that it looks entirely different from virtually all the philosophy that went before. And the differences are differences that matter in your life. Starting with results from cognitive semantics, we discovered a lot that is new about the nature of moral systems, about the ways that we conceptualize the internal structure of the self, even about the nature of truth.

EDGE: This seems like a distinctively new kind of enterprise.

LAKOFF: It's an interesting enterprise to take philosophy as a subject matter for empirical study in cognitive science. Most philosophers take philosophy as an a priori discipline, where no empirical study of the mind, reason, and language is necessary. In the Anglo-American tradition, you are taught to think like a philosopher, and then it is assumed that you can, on the basis of your philosophical training, make pronouncements about any other discipline. Thus, there are branches of philosophy like the Philosophy of Language, the Philosophy of Mind, the Philosophy of Mathematics, and so on. Johnson and I realized that philosophy itself, which consists of systems of thought, needed to be studied from the perspective of the cognitive sciences, especially cognitive semantics, which studies systems of thought empirically. Our goal has been to bring a scientific perspective to philosophy, especially a perspective from the science of mind.

EDGE: How does this connect with traditional philosophy?

LAKOFF: It is a startling thing to realize that most of Western philosophy is inconsistent with fundamental results from the science of the mind. But that is negative. We respect and value philosophy. Our work comes out of a deep love for philosophy and a disappointment over what it has been over the past couple of decades. We wanted to look at great moments in the history of philosophy—the Presocratics, Plato, Aristotle, Descartes, Kant—even the analytic philosophers—and show what shining the light of cognitive science on philosophy could reveal about the nature of philosophy.

What we discovered was fascinating: Each major philosopher seems to take a small number of metaphors as eternal and self-evident truths and then, with rigorous logic and total systematicity, follows out the entailments of those metaphors to their conclusions wherever they lead. They lead to some pretty strange places. Plato's metaphors entail that philosophers should govern the state. Aristotle's metaphors entail that there are four causes and that there cannot be a vacuum. Descartes' metaphors entail that the mind is completely disembodied and that all thought is conscious. Kant's metaphors lead to the conclusions that there is a universal reason and that it dictates universal moral laws. These and other positions taken by those philosophers are not random opinions. They are consequences of taking commonplace metaphors as truths and systematically working out the consequences.

EDGE: What's the import of recognizing that metaphors are central to the work of earlier philosophers?

LAKOFF: It is not just earlier philosophers, but contemporary philosophers as well. Our moral is not that their work should be disregarded because it is metaphorical. Quite the opposite. Because most abstract thought is, and has to be, metaphorical, all rigorous abstract systems of thought will be like those of the great philosophers whose systems of thought we analyze. Moreover, everyone's everyday reasoning is often of the same character, though hardly as consistent overall. A cognitive perspective on philosophy not only teaches us how the great philosophers thought, but it gives us deep insights on how all of us think—at least when we're being consistent and systematic. It also tells us that, in most cases, the answers to the deepest questions of human existence will most likely be metaphorical answers. There is nothing wrong with this. We just need to be aware of just what our metaphors are and what they entail.

Another positive thing we sought to do was to look at the most fundamental of philosophical concepts from the perspective of cognitive semantics. Mark made a list of the basics. In addition to Truth, we looked in detail at Time, Causation, Events, the Mind, the Self, Morality, and Being. Luckily, a fair amount of work had already been done on these within cognitive semantics. We pulled the results together, unified them, and worked out further details. Not surprisingly, all of these abstract concepts turned out to be mostly metaphorical, using multiple metaphors, each with a different logic. Thus, there is not one concept of causation, but around twenty, each metaphorical and each with different inference patterns. Thus, causes can be links, paths, sources, forces, correlations, essences, and so on. Pick a metaphor for causation and different inferences come with the metaphor.

The science and the social sciences all use causal theories, but the metaphors for causation can vary widely, and thus, so can the kinds of causal inferences you can draw. Again, there is nothing wrong with this. You just have to realize that causation is not just one thing. There are many kinds of modes of causation, each with different logical inferences, that physical, social, and cognitive scientists attribute to reality using different metaphors for causation. Again, it is important to know which metaphor for causation you are using. Science cannot be done without metaphors of all sorts, starting with a choice of metaphors for causation. Most interestingly, if you look at the history of philosophy, you will find a considerable number of "theories of causation." When we looked closely at the philosophical theories of causation over the centuries, they all turned out to be one or another of our commonplace metaphors for causation. What philosophers have done is to pick their favorite metaphor for causation and put it forth as an eternal truth.

EDGE: Where does morality come into all this?

LAKOFF: One of the most satisfying set of results is the collection of metaphors governing moral thought. We found that they all seem to arise naturally in an embodied way from forms of well-being—health, wealth, uprightness, light, wholeness, cleanliness, and so on. A particularly interesting result is that moral systems as a whole seem to be organized metaphorically around alternative models of the family. Again, this should not be surprising, since it is in our families that we learn what we take as moral behavior.

We are now in a position to study the metaphorical structure of various moral systems. We think that cognitive science allows one to give much more detailed and insightful analyses of metaphorical systems than has ever been available before. For example, in our study of Kant's moral theory, we argue that this great intellectual edifice arose from just four basic metaphors, and that this allows us to see just how the various aspects of Kant's moral theory fit together.

Cognitive science not only sheds light on the conceptual structure of moral systems, but also on politics and social issues. Some colleagues and I are now in the process of forming a political think tank to apply these methods of cognitive analysis to everyday political and social issues.

Perhaps the most sobering result is the most fundamental. We are neural beings. Our brains take their input from the rest of our bodies. What our bodies are like and how they function in the world thus structures the very concepts we can use to think. We cannot think just anything—only what our embodied brains permit.

Metaphor appears to be a neural mechanism that allows us to adapt the neural systems used in sensory-motor activity to create forms of abstract reason. If this is correct, as it seems to be, our sensory-motor systems thus limit the abstract reasoning that we can perform. Anything we can think or understand is shaped by, made possible by, and limited by our bodies, brains, and our embodied interactions in the world. This is what we have to theorize with. Is it adequate to understand the world scientifically?

There is reason to think that our embodied conceptual resources may not be adequate to all the tasks of science. We take case studies from physics and discuss them in our sections on Time and Causation. General relativity is a good example.

EDGE: So, what's the big change here?

LAKOFF: In characterizing space-time, Einstein, like Newton before him, used the common metaphor that time is a spatial dimension. My present time and location is metaphorically conceptualized as a point in a four-dimensional space, with the present as a point on the time axis. In order for there to be curvature in space-time, the time axis must be extended—it cannot be just one point, the present. In addition to the present, the time axis must include portions of the time axis understood as future and past if there is to be enough of the time axis to form a curved space-time. This seems to imply, as philosophers have repeatedly observed, that at least portions of the future and past coexist with present. And if the future exists at present, then the universe is deterministic. Frankly, it seems nutty to say that the past, present, and future are coexistent—and yet the curvature of space-time seems to imply it.

EDGE: Does the problem lie with the physical theory or the mathematics used to express it?

LAKOFF: It lies with the common metaphor "Time Is a Spatial Dimension," which is used to understand Einstein's mathematical theory of the physical universe. The philosophical entailment of determinism is coming not out of the mathematical physics, but out of that metaphor applied to the mathematical physics. Does that mean that we should—or can—try to jettison the metaphor?

For better or worse, we cannot get rid of it—even if it does have a nutty entailment. Physics is about something. We need to link the mathematics of relativity to an understanding of space and time. The "Time Is a Spatial Dimension" metaphor does that job. We have no better metaphor and no literal concept arising from our embodied minds to replace it with. The commonplace metaphor may be imperfect in having a nutty entailment, but it's the best that embodied human conceptual systems are likely to come up with. What this means is that it is important to separate the mathematical physics from the commonplace metaphors used to comprehend it. And it is vitally important not to take those metaphors literally, even if that leaves us with no literal understanding at all. We should not take time literally to be a spatial dimension; we should recognize that we are using a common metaphor, and that the metaphor has the un-wanted baggage of determinism—the entailment that present, past, and future coexist.

The moral is that you cannot take conceptual systems for granted. They are neither transparent nor simple nor fully literal. From the perspective of the science of mind, science itself looks very different from what we are commonly taught it is. Scientific understanding, like all human understanding, must make use of a conceptual system shaped by our brains and bodies.

George Lakoff

3

Parallel Memories: Putting Emotions Back into the Brain

Joseph Ledoux

Neuroscientist, New York University; Author, The Synaptic Self

EDGE: Emotions and the brain? Isn't this something new for scientists?

JOSEPH LEDOUX: Twenty years ago no one cared about emotions and the brain, but it seems in the last couple of years there's been a flurry of activity. One reason for this may be that the topic was ignored for so long, and the vacuum is being filled. Another, though, is that there have been some successes in approaching the problem, and these have changed people's minds about the feasibility of studying emotions in the brain.

The most successful efforts have come from the study of fear. Fear is a relatively tractable emotion, unlike love or hope, which are difficult to pin down. It's always easier to study brain functions that involve clearly defined stimuli and responses than those that don't. For fear, you can easily create experimental situations where the onset of a simple stimulus that warns of impending danger elicits a set of stereotyped responses in an animal, like a rat, that are very similar to the kinds of responses that occur in a human facing danger. By following the flow of the stimulus through the brain from the stimulus processing pathways to the response control networks, it's possible to identify the basic neural circuits involved. We've done this for fear.

EDGE: How did you get into this?

LEDOUX: I got interested in emotions while I was studying something completely different. I was doing split-brain research as a graduate student with Mike Gazzaniga. Mike and I were studying how information is transferred between the hemispheres of these patients. One of the questions we asked was what happens when we put information in the right hemisphere. Remember, it's the left hemisphere that usually does the talking, so information in the right hemisphere can't ordinarily be talked about in these patients. We put emotional information in the right hemisphere, and the left hemisphere couldn't tell us what it saw, but it could tell us how it felt about it. That led us to the idea that emotional information and information about the content of what a stimulus is are processed by different pathways in the brain. That seemed very interesting, and I decided I wanted to pursue it.

At the time, I felt that the only way to go about studies of the pathways of emotional processing was to turn to an animal model, where you can do experimental lesions, cell recordings, pathway tracing, and so on. The reason you want to do these kinds of studies is not to satisfy some reductionistic urge, but because they can help you see how emotion is put together in the brain, and this can tell you about how the function itself works. Today, there are more sophisticated ways of studying the human brain, such as functional imaging. These can give you a picture of the brain in some emotional state, but you can't then ask the next question. You want to know how the activated region fits into a larger system. You really can't get to those kinds of questions in humans and have to turn to the animal models for answers. The animal work, in other words, gives the framework for interpreting the snapshots we get from human imaging studies. Without the animal studies, though, many of the human studies probably never would have been done, and if they had, they wouldn't be so readily interpretable.

So I left the world of human neuropsychology and went into animal research after finishing my PhD and a short postdoc. Mike and I had moved to Cornell Medical School, and after a year or so I hooked up with Don Reis in the Neurobiology Lab there. The lab's mission was to study the brain's control over the autonomic nervous system, and basically I was told that I could do whatever I wanted as long as I recorded blood pressure. So I developed a blood pressure model of conditioned fear.

I used conditioned fear because it seemed like a relatively straight-forward technique: You give a meaningless tone followed by a mild shock a few times, and pretty soon the tone starts eliciting a blood pressure response. It was a good way to create an emotional reaction to the tone on the spot in an animal that wasn't afraid of the tone and didn't have any emotional reaction to it to begin with. Since the tone gets to the brain by way of the auditory system and the response comes out of the brain through the autonomic nervous system, the trick was to figure out how the auditory system is linked up with the autonomic system. By using a combination of brain lesions, neural recordings, and pathway tracing techniques, we were able to figure this out. The answer, in short, is that the amygdala turned out to be a necessary and sufficient link between the auditory system and the autonomic nervous system. However, in a more general sense, the amygdala is the link between all sensory systems and all fear response systems. It's the part of the brain involved independent of how the stimulus gets into the brain and how the response comes out.

EDGE: I find it interesting that the first emotion you studied was fear.

LEDOUX: When I first began this work in the early 1980s, I was using fear conditioning techniques because they were convenient. As I said, you can take the stimulus, pair it with the shock one or two times, and, as a result, create an emotional reaction that's relatively profound in the animal. I thought at the time that this was going to be a way of identifying a universal emotional system in the brain, something akin to the limbic system. I no longer feel that way. I think that the study of the limbic system, or more generally the idea that there is an emotion system in the brain, is misguided. I came to this conclusion empirically. Once we had outlined a neural circuit for fear responses, it was obvious that the limbic system had little to do with it. The only so-called limbic area involved was the amygdala. And the hippocampus, the centerpiece of the limbic system, had been implicated in nonemotional processes like memory and spatial behavior. It seemed clear that the limbic system, if it existed at all, was not systematically involved in any clear way. I decided I didn't need the limbic system concept to think about how fear works in the brain. But that still doesn't wholly justify the focus on fear to the exclusion of other emotions.

I've come to think that emotions are products of different systems, each of which evolved to take care of problems of survival, like defending against danger, finding mates and food, and so forth. These systems solve behavioral problems of survival. Detecting and responding to danger requires different kinds of sensory and cognitive processes, and different kinds of motor outputs, different kinds of feedback networks, and so on, than finding a mate or finding food. Because of these unique requirements, I think different systems of the brain are going to be involved in the different kinds of emotions.

A related point is that emotion systems, like the fear system, didn't come about to create feelings (like the feeling of being afraid when in danger). I think feelings came much later in evolution. All animals have to be able to detect and respond to danger, regardless of the kind of cognitive architecture they have. This is as true of bees and worms and snails as it is of fish, frogs, birds, rats, and people. Fear conditioning, by the way, occurs in all animals. And in all those that have an amygdala, the amygdala appears to be the key. The list at this point includes reptiles, birds, and a host of mammals, including humans. I think it's safe to say fear behavior preceded fear feelings in evolution. If so, feelings are probably the wrong thing to focus on when we

Joseph Ledoux

study emotions. In this sense, animals were unconscious, unfeeling, and nonlinguistic before they were conscious, feeling, and linguistic. It's too bad that we define the more basic processes as the negation processes that typify the human brain. It's possible that once consciousness and feelings came along that new kinds of emotions specifically tied to these evolved. But I'm trying to understand the things about emotions that are similar in humans and other animals so that I can work on emotions through the brain.

I tend to agree with theorists who say there are basic emotions that are hard-wired into the brain's architecture, and that one of the advantages of having an extra-big cortex is that we can blend different hard-wired emotions together to create softer emotions, where cognitions come into play in a major way. For example, while detection and responding to danger may be built into the brain, the capacity to be afraid of falling in love is something that requires the cognitive integration of the system for finding mates and the system for defending against predators. While I'm sympathetic toward the basic emotions view, I don't really ascribe to it. It requires that you state what the different emotions are. That just leads to arguments. I'd rather spend my time worrying about one well-accepted emotion and its organization in the brain than fighting over whether this or that mental process is an emotion or not.

EDGE: So what about feelings? What are they?

LEDOUX: The study of emotion has focused on conscious feelings almost to the exclusion of everything else. Emotion researchers, for some reason, seem to be carrying the burden of the mind-body problem on their shoulders. In other words, I think the problem of feelings is one and the same as the problem of consciousness, and that emotion researchers have no more or less of an obligation to solve this before anybody else. Take vision. Philosophers have worried about where the redness we experience comes from when we see an apple. But vision researchers figured out that they could study how we process red without having to first figure out how we experience red. The same can be done in the study of emotion. We can study how the brain detects and responds to danger, even if we don't know how it experiences danger. So the feelings of fear that come about in dangerous situations are in a sense no different from any other kind of conscious experience. The only difference is in the system that consciousness is paying attention to—the danger processing system, the color processing system, the language processing system, and so on. So emotional feelings come about when we become consciously aware of the activity of an emotional system, which does its work for the most part outside of consciousness.

EDGE: What's the difference between an emotional and a cognitive memory?

LEDOUX: By cognitive memory I'm going to assume you mean explicit conscious memory, the kind of memory we usually have in mind when we use the word "memory" in everyday speech. Emotional memory and explicit memory happen at the same time, but separately. For example, the amygdala mediates emotional memory and the temporal lobe memory system mediates explicit memory.

Here's an example. Imagine driving down the road and having an accident. You hit your head on the steering wheel and the horn gets stuck on. You're bleeding and in pain. It's awful. Sometime later, you hear the sound of a horn. The sound goes to your amygdala and activates your autonomic nervous system (raising your blood pressure and heart rate, making you sweat), tenses your body muscles, releases stress hormones into your blood, and so on. The sound also goes to the temporal lobe system and reminds you of the accident, of who you were with and where you were going. It also reminds you that it was awful. But these are all just facts about the situation. They are memories of the emotional experience rather than emotional memories. In general, one difference between emotional and cognitive processing is that emotional processing often leads to bodily responses, whereas cognitive processing leads to more cognitive processing. Cognitions are seldom characterized by specific kinds of responses, but emotions usually are. It's important that we understand as much as we can about the biology of these systems.

Many people have problems with their emotional memories; psychologists' offices are filled with people who are basically trying to take care of and alter emotional memories, get rid of them, hold them in check. If anything, emotional memory is more basic than explicit conscious memory. For example, it takes place at an earlier age. It's conceivable, and in fact seems very likely, that a child could be abused very early in life and develop unconscious emotional memories through the amygdala prior to the point where the temporal lobe memory system has kicked in. If that's true then emotional memories are being formed for things that will never be consciously understood, because the system that mediates conscious memory isn't available to encode the experience and can therefore never retrieve it.

We need to understand how unconscious emotional memories are formed—not only because they occur in early childhood, but because emotional memories are created throughout our lives. And it appears that these memories are indelible. They can be extinguished in the laboratory or treated in the psychiatrist's office, but they can usually be brought back. And recently we've been able to find a mechanism in the amygdala that might be responsible for this. It's sort of complicated, but the finding goes like this: We record neural activity in the amygdala before and after conditioning. Cells fire more to the tone afterward. With extinction the firing rate goes back to baseline. However, in addition to measuring these stimulus-evoked responses, we measure the correlation in the time when different cells fire spontaneously (no stimulus present). After conditioning, some cells that were not correlated become correlated. And for some of the cells the correlations remain past extinction. In other words, the feared stimulus no longer elicits activity in the amygdala, but the amygdala cells continue to be functionally coupled. It's as if extinction (and therapy) doesn't erase the memory, it just weakens the ability of the stimulus to activate the memory. So in order for the stimulus to again be effective, all you have to do is change the synaptic strength of the connection between the stimulus and the memory rather than recreate the memory.

This is relevant to phobia, where the phobia can be in remission (the sight of a snake no longer elicits paralyzing anxiety), and then the patient's mother dies and snakes regain their propensity for producing terror. Phobia is also a good way to illustrate the difference between cognitive memory of emotion and emotional memory. We aren't born with phobias. Somehow they are acquired through experience and stored in the brain as links between stimuli (like snakes or heights) and fear responses. Once a phobia is successfully treated (the snake no longer elicits overt fear responses) the patient still has the explicit memory of having had the snake phobia. In other words, the therapy inhibited the amygdala's pathological response to the sight of snakes, but the therapy didn't eliminate the temporal lobe memory system's memory of having had a snake phobia.

EDGE: How can you talk about unconscious emotional memories? Why is this different than inventing a concept like "repressed memories"?

LEDOUX: I'm not talking about memories that have been repressed, they're just not consciously available. I'll give you a simple example. Patients who have damage to the temporal lobe memory system are unable to remember what happened to them five minutes ago. If you take those patients and give them a sound, pair it with a shock, and you later give them the sound again, their autonomic nervous system responds, but they have no conscious memory of the experience that led to that. The memory is in the brain, having an effect on systems that we can measure, including autonomic and behavioral systems, but the patient has no conscious memory of it. In everyday usage, the term "memory" usually refers to conscious memory, but as scientists we use the term in a more general sense to mean changes in the nervous system that reflect past experiences. By this definition, we can see all sorts of memories that have no conscious counterpart. This is the idea of implicit, or procedural, memories that are in the brain's systems, but not reflected in consciousness.

There's a famous case from the early days of the 20th century that beautifully illustrates this point. The patient had a pretty severe amnesia. Each day she had to be reintroduced to her doctor, as she didn't recognize him. One day the doctor put a tack in his hand, and he walked in and shook her hand. When their hands met, her finger was pricked. He then walked out of the room, walked back in, and asked whether she'd ever seen him before. She said she hadn't. But when he stuck out his hand to shake hers, she held back. Although we don't really know what was going on in her brain, it seems likely that the implicit memory that the handshake was dangerous was burned into her amygdala, and that allowed her to protect herself from getting stuck again. She knew this implicitly—but she couldn't tell you why because she couldn't remember the experience that led to it. The amygdala was forming its memories, but the temporal lobe memory system was not.

Normally, these systems work in parallel to give rise to our conscious memories about emotional experiences, and unconscious emotional memories. In this sense, emotional memories are by definition unconscious. But they aren't unconscious because they've been repressed. They're unconscious because they are not formed by the conscious memory system. The conscious memory system forms memories about emotions, but doesn't form the emotional memories that have direct access to emotional response systems.

EDGE: Where do you expect your research will take you?

LEDOUX: Right now my work is headed deeper and deeper into cellular-molecular events underlying how emotions are learned and stored. We are trying to understand as much as we can about how these memories are formed at the cellular level, which has taken us into studies of synaptic plasticity, how changes happen at the level of individual synapses when this kind of learning takes place. We are asking questions about what neurotransmitters are involved and what sort of molecular changes take place to stabilize these memories over the long run. These studies are just beginning and they will take us well into the 21st century. At the same time it's important not to lose sight of the fact that we're dealing with a psychological problem with important behavioral consequences. We need to study the behavior as well as the molecules. We try to work at all these levels, at the level of the behavioral system as well as the cellular and molecular systems.

EDGE: What has your work made explicit?

LEDOUX: There are ways that the brain can produce emotional responses in us that have very little to do with what we think we're dealing with or talking about or thinking about at the time. In other words, emotional reactions can be elicited independent of our conscious thought processes. For example, we've found pathways that take information into the amygdala without first going through the neocortex, which is where you need to process it in order to figure out exactly what it is and be conscious of it. So, emotions can be and, in fact, probably are mostly processed at an unconscious level. We become conscious and aware of all this after the fact. Conscious feelings of fear are thus not a necessary step in the link between a dangerous stimulus and emotional responses. We're probably not as in control of our emotions as we sometimes think we are, or wish to be.

Emotional reactions that occur in this quick and dirty way are really reactions that are important in survival situations. The advantage is that by allowing evolution to do the thinking for you at first, you basically buy the time that you need to think about the situation and do the most reasonable thing. For example, freezing is often the first thing people and other animals do when sudden danger appears.

Joseph Ledoux

Predators respond to movement, so freezing is overall probably the best single thing to do first; at least it was for our distant ancestors. If they had to think about what to do first, they'd have been so caught up in the thought process they'd probably fidget around and then get eaten.

The Atlanta Olympic bombing is a nice illustration of this. The bomb goes off and everyone hunches over in the freezing posture for a couple of seconds, and then they take off running. You can almost see the cognitive gears turning while they're freezing. Although we're not in direct control of these rapid-fire unconscious emotional responses, I don't think that they are necessarily going to be things that someone can use as a legal defense, for example, for having carried out a very detailed crime, a murder or a rape or something of that nature. These quick and dirty systems produce relatively simple rapid responses (like freezing) in life-threatening situations. They're more likely to be used by the victim than by the perpetrator.

EDGE: What about therapy?

LEDOUX: The connectivity of the amygdala with the neocortex is not symmetrical. The amygdala projects back to the neocortex in a much stronger sense than the neocortex projects to the amygdala. David Amaral has made this point from studies of primate brains. The implication is that the ability of the amygdala to control the cortex is greater than the ability of the cortex to control the amygdala. And this may explain why it's so hard for us to will away anxiety; emotions, once they're set into play, are very difficult to turn off. Hormones and other long-acting substances are released in the body during emotions. These return to the brain and tend to lock you into the state you're in at the time. Once you're in that state it's very difficult for the cortex to find a way of working its way down to the amygdala and shutting it off.

This is why therapy is probably such a long and difficult process, because the neocortex is using imperfect channels of communication to try and grab hold of the amygdala and control it. It's like trying to find your way from New York to Boston by way of country roads rather than superhighways. The amygdala can control the neocortex very easily, because all it has to do is arouse lots of areas in a very nonspecific way. But for the cortex to then turn all of that off is a very difficult job. The evolution of the brain is at a point where we don't have the connectivity that would be necessary for cognitive systems to more efficiently control our emotions. But it's not clear to me that would necessarily be a good thing, because Mr. Spock is not necessarily an ideal kind of human that we'd like to become.

Designer drugs could be really practical, and I'm surprised that the drug companies are not knocking at my door to find out how to make drugs that could do more specific things than the drugs that are available. We know the circuit through which fear is elicited, and we know the specific points in that circuit that are involved. As we begin to identify the neurotransmitters that are involved in the elicitation of fear, it seems that we could probably come up with a chemical profile of fear in the amygdala. A particular drug could be developed that attacks that profile. For example, if you take Valium, it might make you sleepy and reduce your sex drive in addition to making you less anxious, because it affects GABA transmission throughout your entire brain. But if you could develop a Valium that only acted in the amygdala, then you would have a drug that works at the particular sites involved in fear. That's pie in the sky at this point, but it's something they should be thinking about.

EDGE: How is your work being received today?

LEDOUX: I've been amazed that almost all areas of psychology have not only been sympathetic, but are reaching out and trying to find out as much as possible about my work and the work of people like me. It's really surprising that this extends into psychoanalysis as well. I have received a number of invitations to speak to psychoanalytic groups and to attend meetings to help them understand concepts about the emotional brain and that psychoanalysis has to change as we enter the 21st century. Psychoanalysis is in relatively bad shape right now. Young psychiatrists are not going into the field, so the elders are trying to figure out a way to make the field more appealing. I think they see neuroscience as a possible bridge.

EDGE: Who else?

LEDOUX: Developmental psychologists and social psychologists have been very open to the work on the emotional brain. The developmental psychologists are interested because of the early development of the amygdala before conscious memories kick into play. Social psychologists are interested because the amygdala seems to do its work unconsciously. There's a whole industry of social psychology dealing with unconscious emotional perception, how you use subtle cues that are given off even when you don't know you're giving them off, and how these are picked up by your unconscious mind, so your unconscious mind and my unconscious mind are talking back and forth to each other without our conscious minds knowing anything about it. They're interested in all this work on the amygdala and the possibility that it's an unconscious emotional processor.

Cognitive scientists previously banned emotion from their field, but are beginning to realize that they don't really have a science of mind as such, but instead a science of a part of the mind. They now want to bring emotion and cognition back together, and that's a good thing. Lots of AI modeling of emotion, and some connectionist modeling, is also going on.

EDGE: How does a Dan Dennett or a Steve Pinker relate to your work?

LEDOUX: In Pinker's talk, "Organs of Computation," I noticed that he did talk about emotions; he was talking about how passions fit into the mind. I think we'd agree on a lot, say, about the evolutionary aspects of emotions and their unconscious nature as processes in the mind. I'm more interested in how evolution has kept emotional systems the same in man and other animals, whereas he seems to be more interested in what makes the human brain capacity for language a unique function not present in other species. Where we'd probably differ the most is in how we approach the problem. I want to do it from the brain, so that I know that my theories are tied to the hardware in a biologically plausible way, but I think he wants to do it without depending on the brain. I think both approaches have their strengths and weaknesses, and both are needed.

What I talk about is compatible with Dennett's views in some ways, because I'm dealing with emotions not as conscious feelings but instead as computational functions of the nervous system. The way I talk about emotions puts them at the level of what some people in cognitive science call the sub-symbolic level. In this sense, emotional systems are among many systems that operate in parallel at an unconscious level. In Dennett's view, there's a symbolic system sitting on top of all these sub-symbolic systems. This is where consciousness comes from, loosely speaking.

The symbolic system has some access to the outputs of the unconscious emotion systems as well as all the other perceptual and other sub-symbolic systems, and the one that grabs hold of the symbolic system at the moment is what we are conscious of. So we can be conscious of emotional events or mundane events. So I'd say there's not a special system for emotional experience separate from other kinds of conscious experiences. There's one mechanism of consciousness and it can be occupied with mundane events or highly charged emotions. I think my view of the mind is not incompatible with Dennett's. That's not to say that I agree that Dennett's explained consciousness. Instead, I agree that most of the mind doesn't work through consciousness.

EDGE: How would you describe yourself?

LEDOUX: I was recently called a radical behaviorist disguised as a neuroscientist. I thought that was an interesting twist. It's true that

I try to deal with emotions as unconscious processes as far as I can, but I don't deny the importance of consciousness. I just think that it's gotten in the way in the study of emotions. I'm not really a radical behaviorist. I realize that I am simplifying and probably oversimplifying emotion to study it the way I do, but I hope to build up to complex issues from a solid understanding of the simple stuff rather than have to reach down from confusion to try and account for the simpler processes.

EDGE: Can you say more regarding the difference between repressed memory and a sub-symbolic system?

LEDOUX: There are several things that are important to pull into this topic. One is the newly emerging data on the effects of stress on memory. The basic finding is that in periods of intense stress the explicit memory functions of the temporal lobe memory system can break down. Stress is usually defined physiologically by the amount of so-called stress hormones from the adrenal gland. When this stuff is released it floats around in your bloodstream and gets into the brain. The hippocampus and amygdala are targets. These hormones adversely affect the hippocampus. They make it very difficult, for example, to induce long-term potentiation in the hippocampus, so the hippocampus begins to shut down physiologically. Also, spatial learning is interfered with. If the stress continues, dendrites begin to shrivel up, and if the stress continues even longer the cells die and the hippocampus itself begins to shrink in size. Bruce McEwen and Robert Sapolsky have done a lot of this work on stress and the hippocampus. There have also been some recent studies of patients with posttraumatic stress disorder—Vietnam vets, for example, who have a greatly reduced volume of their hippocampus, and they have all these memory disturbances.

In contrast, stress seems to potentiate the amygdala. Stress will make the amygdala do what it's doing but even better. Let's say you get mugged or raped. The stress system releases all of its hormones (probably as a result of the amygdala detecting the threat and activating the stress hormone system). The hormones get into the brain and the hippocampus is adversely affected to the point where it can't consciously form a memory of this experience. But your amygdala is potentiated, so it's not only forming a memory unconsciously, but it's doing it better than before. So the exact conditions that can lead to hippocampal memory impairment (an inability to form a conscious memory) can lead to a facilitation of unconscious emotional memories through the amygdala.

Now you're a person who's walking down the street with no conscious memory of having been traumatized. There are witnesses that tell you it happened but you deny it—there is, in fact, often denial in situations like this. You carry unconscious traumatic memories but no conscious understanding of what happened. I don't know that something like this actually happens, but the biology is very plausible. It's totally conceivable that someone can be traumatized in this way and have no conscious memory of it. I believe that. And it fits with all the science that we have about all of this.

Now the next question is, can you then, through psychological tricks, comforting, therapy, whatever, bring these memories back in a person who never had them? And the answer to that is a clear No. It's not possible to take a memory that was not coded through the hippocampus and turn it into a hippocampal memory. So the amygdala has its memory; it doesn't then share it with the hippocampus, because they do things differently. The amygdala does its business, the hippocampus does its business. They communicate with each other, but their coding and representation is different. So you can't just get information out of the amygdala and turn it into content that the hippocampus can read. I think this kind of work tells us a lot about the psychology of memory and emotion, not just the biological details.

EDGE: What do you want to accomplish in the next five years?

LEDOUX: I want to understand several aspects of emotion that we have very poor understanding of now. The first part we're beginning to understand pretty well, which is how the initial aspect of an emotional reaction is elicited. In other words, how you jump back from a bus as it's approaching, and only afterward consciously realize that you've jumped back, and only then feel afraid. We understand that reactive system in pretty good detail. But what we don't understand is the system for emotional action. How do you voluntarily make decisions and control your emotional behavior once you've reacted in this unconscious way. What circuits in the brain are involved in what psychologists call coping, the cognitive and behavioral effects that follow the arousal of emotion and one's attempts to deal with emotional arousal? Probably the basal ganglia and cortex are involved. The question of what makes us emotional actors as well as reactors really interests me.

That takes us to another issue, which is where do conscious feelings come into emotions? How do we get a deeper understanding of emotional feelings? We all want to know where feelings come from and how they work. So much of the work in the past started with feelings and tried to back into the problem and didn't get anywhere, which is why I start at the bottom and work up to feelings. I also want to know a lot more about emotional memory. Most of the things that make us emotional are learned through experience. So a key part of an emotion system is how it learns and stores information. Overall, I would summarize all this by simply saying I want to try to understand more about cognitive-emotional interactions. We have to put emotion back into the brain and integrate it with cognitive systems. We shouldn't study emotion or cognition in isolation, but should study both as aspects of the mind in its brain.

4

Sexual Selection and the Mind

Geoffrey Miller

Evolutionary Psychologist, University of New Mexico; Author, The Mating
Mind *and* Spent: Sex, Evolution, and Consumer Behavior

My goal at this point really is to take evolutionary psychology to the
next step, and to apply standard evolutionary theory as much as possible to explain the whole gamut of the human mind, human emotions,
human social life, human sexual behavior. I'm especially interested
in looking at areas that have been relatively ignored or overlooked in
the standard evolutionary psychology so far. For example, in Steve
Pinker's book *How the Mind Works* there's a very good discussion of
vision, memory, emotions—but some of the most interesting aspects
of the human mind, such as art, music, humor, and religion, tend to
get relatively slighted, and it's apparent that we don't have very good
explanations of them yet. I'm very interested in applying sexual selection ideas to explain some of those areas, but I'm quite open to any
new ideas that come along that take seriously those aspects of human
nature that have not been taken seriously before.

Another thing I'm interested in at the moment is trying to create
more cooperation between evolutionary psychology and behavior
genetics, especially for understanding the mind, and distinguishing
between parts of the mind that are truly universal, where everybody's
got the same structure, versus parts of the mind where there's significant variability between people, and where some of that variability is
genetic. There's been too much hostility between behavior genetics
and evolutionary psychology, too much mutual misunderstanding.

Evolutionary psychology is studying human universals. Structures and adaptations in our minds where everybody's got the same stuff, everybody has the same abilities. Behavior genetics is traditionally studying differences between people—for example, differences in intelligence, differences in personality. And, their aims have been different.

Behavior genetics tries to figure out, are the differences between people due to genetic differences or environment differences? And so far the answer seems to be, as far as we can tell, surprisingly, the genetic differences are very powerful. And evolutionary psychology just hasn't coped with this news yet, and is not making the best use of powerful new DNA research methods from genetics. And there's serious unresolved questions about the nature of human intelligence itself. We know from intelligence research that there is such a thing as general intelligence—there's a G factor—people differ on this dimension that accounts for hugely important things, like success in education and real life, and we know that people who tend to be good at certain kinds of mental things like having a large vocabulary also tend to be good at other mental tasks such as mathematics, or spatial navigation. Why are there these correlations between mental abilities?

People have the mistaken idea that general intelligence, as it's talked about in intelligence research, is somehow contrary to the views of evolutionary psychology. The evolutionists say our minds are a collection of different capacities, different adaptations for doing different things. And from that point of view there's no such thing as a general intelligence that spans these capacities, or that sits on top of them directing everything. That does not actually conflict with what intelligence researchers think. The best intelligence researchers admit there's no such faculty as a general intelligence; this G factor is just a statistical abstraction, it just captures the ways that people who are good at one thing are also good at another thing. There's a fight now between intelligence research and evolutionary psychology that doesn't have to exist. It's easy to solve; I hope we can have a conference about it soon, and both fields will be better off for resolving that confusion.

Some of the leading evolutionary psychologists, like Steve Pinker, Leda Cosmides, and John Tooby, have a very good understanding of the mind's architecture, but they sometimes don't seem up-to-date about individual differences and about intelligence research.

On the other side, some of the leading intelligence researchers, like Arthur Jensen, Ian Deary, and Robert Plomin, understand that the mind might be a collection of different capacities, but they're also starting to find powerful indications that different people have different brains that operate at different degrees of efficiency, and some of those efficiency differences are due to genetic differences.

The study of human intelligence is really explosive, ideologically, politically, and socially. It was a good strategy for the early evolutionary psychologist to distance himself as much as possible from genetics, from individual differences, from the study of intelligence, because they could avoid all of this political firestorm surrounding those issues, and they could get on with the job of describing human nature, where it comes from, how it works, why it's there. But evolutionary psychology is now established, and we don't have to make the same mistakes and we don't have to be as cautious and shy about avoiding some of these controversial issues.

There are serious considerations, serious downsides to studying individual differences. Evolutionary psychology has been successful in teaching people how to think properly about sex differences already, which used to be a really contentious area; now people are starting to cope positively with the idea that there might be important differences between male and female psychologies, especially in terms of social and sexual behavior—and that used to be a totally taboo area, that used to be outlawed, to talk about evolutionary or genetic differences between the sexes. People's sophistication about the sex difference issues is starting to catch up with the sciences.

In the area of individual differences it's going to be a really hard fight to teach the general public the concepts and attitudes that they need to properly understand research in this area—to properly understand the genetics.

One thing to understand is some of the positive sides, especially for parents, of understanding the importance of genetic inheritance and individual differences: Now, toy manufacturers and purveyors of educational materials are doing a very good job of convincing parents that they have to give their kids the optimal environment for intellectual growth, and that if they don't spend the money to do that, on—the money spent on toys, on the right child care, on the right private schools, the right universities—then the child's going to be a failure. And that if they don't push their child and motivate them, and be worried for the entire time that they're growing up about what they're going to become, then they're going to be failures. That's completely the wrong attitude, and Robert Plomin has been very good at pointing out that the more you understand about genetics, the more you can just relax and love your kids for who they are, and who they turn out to be, and the interests that they show, and you can abandon this idea that the kids are born as formless blobs and you have to shape all of their desires and their capacities yourself. It removes some of the burden and anxiety from parents.

Also, for educational policy, understanding individual differences is absolutely crucial. In Britain we have things called league tables for ranking high schools. They rank them by the outcome of high school exams called A levels. Always the private schools that cost the most come out at the top of the league tables. Of course, they might come out because they're taking in brighter students, and the brighter students do well and it has nothing to do with the quality of the teaching.

To properly measure the quality of education, the quality of teaching, you have to measure what the students are like when they come into a school and then what they're like when they go out. You have to have a value-added measure. The only way to do that is to have some good tests of their capacities and their knowledge when they come in. At the moment nobody's doing that in Britain, and very few people are doing it in America. It's going to be difficult for people to cope with ideas that there are just a few measures that can describe—not just their intelligence but their personalities, and that some of those differences might be relatively stable across their lifetimes, and relatively hard to change. But look, in a sense we all know this already, and we especially know it for other traits, like physical attractiveness, and height. Kids are growing up and they sort themselves out into little social hierarchies based on all kinds of things, and we all have to learn to cope with the traits and abilities we have—how physically attractive we are, how tall we are, how athletic we are, as well as what our intellectual capacities are, and what our personalities are. It will be nothing new coping with this new marriage between evolutionary psychology and the new genetic research. It's just a matter of learning to be realistic about ourselves in an area where we've been allowed to get away with wishful thinking for a long time.

In the matter of sex, it's extraordinary what's been happening in biology, and so few people in the social sciences know about it. Over a century ago Darwin's idea of sexual selection through mate choice was published in his best book, *The Descent of Man and Selection in Relation to Sex*—that was the full title—the book came out and this wonderful idea of female choice—the idea that female animals of many species choose their mates for all kinds of traits, not just physical appearance, but behavioral traits, songs, and dances, and courtship behaviors. A wonderful scientific theory that Darwin advanced hundreds of pages of evidence for, and it fell like a stone and was widely rejected by Victorian biologists, who refused to believe that this psychological process of female choice could be a causal force in evolution.

This theory of mate choice languished in a sort of scientific limbo for over a century, and it's only been revived in biology in the last fifteen years, but its rise has been meteoric; it dominates the best evolution journals, the best animal behavior journals, and everybody who works in a biology department knows that the study of mate choice is now the hottest topic in the study of animal behavior. This revolution has passed by psychology and social science almost completely. All of psychology, anthropology, the humanities, political science, economics in the 20th century developed without any understanding of how sexual selection could have shaped human behavior.

It was just not on the table as an idea. Everything that we are, every aspect of human nature, had to be explained through survival selection—natural selection. And that imposed such serious restrictions on what we could explain—it seemed easy to explain tool making; it seemed hard to explain music. It seemed easy to explain parenting, but hard to explain courtship. All that's changed now. We've got from biology some powerful new principles about sexual selection that are just ripe for applying to human nature. That's what I'm trying to do; lots of other people are doing it as well, and it's the most exciting area to be working on in psychology at the moment.

This revival of sexual selection in biology was of course promoted very strongly by people like George Williams, E. O. Wilson, an ingenious Israeli biologist, Amotz Zahavi, and many theoreticians working alone in their offices writing down mathematical proofs showing that sexual selection could indeed work, just the way Darwin thought.

Some of the exciting new ideas coming out now are that many of the traits we're selecting when we choose a mate are not just arbitrary traits, they're not random, they're not meaningless, but they're actually powerful indicators of things that matter in reproduction—that a lot of beauty is really an indicator of health and fertility, and a lot of traits that are psychologically attractive to us, like kindness, warmth, creativity, intelligence, and imagination, also are not random but actually are indicators of somebody's ability to get along in the world—not just the physical world but the social world, and that in choosing

Geoffrey Miller

a mate for these psychological qualities, we're insuring that we have a partner with whom we can have a constructive relationship, rear successful offspring, and pass their better-than-average genes on to our children. What we're seeing here is in studying how people choose their mates, not just for physical appearance but for all these rich psychological traits, it's a wonderful confluence between evolutionary biology, personality theory, and evolutionary psychology. And that to me is very exciting.

One of the great surprises for David Buss, one of the leading evolutionary psychologists studying mate choice, was that when he did his wonderful study in the late 1980s of sexual preferences in thirty-seven cultures all around the world, giving questionnaires to 16,000 subjects that just span all sorts of cultures with all sorts of languages with different traditions and different histories, he found that in every culture, the top two most desired traits in a mate, for both sexes, were kindness and intelligence. It wasn't physical appearance, it wasn't money, it wasn't status, it was these psychological traits, and these are universally important. They're also the two traits that Darwin tried to explain about our species—why are we so nice to each other and why are we so smart? Relatively nice, compared to other primates. And that's fascinating, that two of the major traits that distinguish us from other primates are the same traits that we search for in mates—that are currently under the strongest sexual selection. My hypothesis is that they're not just under sexual selection now, but they have been for a very long time, perhaps hundreds of thousands of years, and the reason we're so smart and so relatively kind to each other is that our ancestors who were smarter than average and kinder than average attracted more mates and higher-quality mates.

Another interesting question is about language. Language is a really tricky case because, as Steven Pinker has pointed out, language is extremely useful for many functions. You can tell your friends how you're going to hunt an animal and cooperate on tracking it down.

Women can tell their friends where the best roots and berries and tubers are growing this season. Parents can tell their kids all sorts of useful information as they're growing up. Of course the principal way that people court each other is through language. Human courtship is largely through conversations. It would be foolish to say that sexual selection was the only force shaping language; clearly survival selection and many other forces were shaping it as well, but I would claim that some of the more mysterious aspects of language can be understood only by thinking about how language is used in courtship.

A project I'm very excited about at the moment is trying to understand why humans have such large vocabularies. The average human knows about a hundred thousand words by adulthood. That requires memorizing arbitrary patterns that relate sound to meaning. It requires memorizing ten words a day every day from age eighteen months to age eighteen years, and that's a fantastic feat of learning. There's nothing else like it that humans do. The funny thing about that vocabulary is how little of it we use in ordinary conversation. We get by in our day-to-day speech with just a few thousand words counting for 95 percent of all the words that we say.

There's a tremendous number of words we've learned that are not used very often but that we bothered to memorize, that don't seem to be very useful in ordinary day-to-day life, but that we still sometimes use with each other—and those are the words that I want to explain—not the 5,000 most useful words but the 95,000 ornamental words. My prediction is, people mostly use them in courtship. They use them essentially to show off; they use them to show how bright they are, how good their learning ability is, how good their memories are for words. We know in the brain where these words are remembered, roughly in Wernicke's area, in certain parts of the left hemisphere; we know that there's specialized brain machinery for learning these words; we know that vocabulary size is an extremely powerful indicator of intelligence—this is why vocabulary items are used in IQ tests; within a few minutes of conversing with somebody you use the vocabulary that they're producing as a pretty good indicator of how intelligent they are—so it's an extremely useful thing to use in mate choice. The hypothesis here is that vocabulary size itself has been strongly shaped by sexual selection, and that most of the words that we know have been learned not because they're useful for survival, but because they're useful for courtship.

Another mystery is why we enjoy music so much—and this is one of the questions that Nicholas Humphrey has asked in the *Edge* forum. Music has such powerful emotional impact, and nobody has ever found a good survival function for it. The very first serious conference on the evolution of music only took place last year, at a wonderful little town called Fiesole, Italy, in the hills overlooking Florence. It became abundantly clear to me at this conference that there were amazing parallels between human music and birdsong, and whale song, and all the other complicated acoustic signals that animals send to each other—even gibbon song. The most musical apes are gibbons who do wonderful duets, long calls that they give to each other, especially to their sexual mates. Wherever you're looking in nature, if an animal is producing a complex acoustic signal it's almost certainly a courtship signal, it's almost certainly involved in sexual selection. We know this for birdsong, we know it for whale song, and we know it for gibbon song.

Darwin thought the same should apply to human music, that human music was largely an outcome of courtship displays. That's a wonderful overlooked theory, and it's surprising that people have scrambled for a century, coming up with all kinds of silly hypotheses about music functioning to make people in a group feel closer to each other and to facilitate group cooperation, for example—that's a favorite idea. If you go to any nightclub in London or New York or Berlin or Tokyo you can see the proper context for understanding music's function. Although it's done in groups the point of it is individual display.

Music combines exactly the features that an evolutionary biologist would predict—for something that indicates an individual's creativity, motor control, self-confidence, and lots of other traits that are important in courtship. Music is a system of basic elements, notes, that are combined according to certain principles of rhythm, tonality, and we know that the basic principles of rhythm and tonality and melody are universal and cross cultures. Even though many of the musical styles are different. And that people can demonstrate their coordination and virtuosity, both as musicians and dancers, by using this system that has stereotyped basic units. The essential thing about rhythm is that you can see whether somebody is rhythmic or not, whether they're coordinated or not. If rhythm didn't exist it would be hard to tell whether somebody was keeping to a regular beat, and whether they could coordinate their body and their musical productions according to a regular beat.

To tell how good somebody is at something there have to be some rules, there have to be some regularities, but for them to demonstrate how creative they are, how innovative they are, they also have to be able to play around with those basic elements, and play around with those rules. Music also provides great scope for that—for melodic innovation, for improvisation, for producing innovative lyrics, for producing unusual timbres when you're singing, or playing an instrument. It's the perfect display, really, for sexual selection theorists. Art and language and many other display forms that we have follow some of the same rules—we combine basic elements that are stereotyped in ways that are innovative, and that's a recipe that you need to indicate your quality to a sexual prospect.

It's fine to talk about all these just-so stories, these evolutionary hypotheses, about why this evolved, why that evolved. Evolutionary psychology is getting much more sophisticated about the methods it

Geoffrey Miller

uses, experiments and observations, to test some of these theories; the wonderful thing about mate choice is that there are already a large number of methods that biologists use routinely to study animal mate choice that are just starting to be applied to human mate choice. But equally important, there are a lot of methods for studying these courtship displays themselves—to see whether their features and how their features indicate the quality of the person producing them.

What I want to do next is really try to flesh out my hypotheses about art and music and language and ideology as courtship displays, to see if they really have the necessary features to really indicate the things about a sexual prospect that need indicating. This is going to require basically measuring lots of correlations—seeing, is vocabulary size really a good indicator of intelligence? Is it a costly display that indicates your quality? Is it noticed, do people pay attention to it? The reason why I'm trying to get my ideas better known in evolutionary psychology and among the general public now is that testing big hypotheses like this is too large a job for any one individual to do—it requires cooperation between dozens or hundreds of people. It took one person to think up Darwin's sexual selection idea, but it took hundreds and hundreds of theorists and animal experimenters to actually show that his theory works. The same is true of trying to apply Darwin's sexual selection ideas to understand human nature.

Unfortunately there are a great number of biologists who shy away from applying evolutionary theory to the human mind. A large part of it is a failure of nerves—that they're comfortable getting grants to do research on animals, and those grants might be threatened or compromised if the public understood that the theory that they're using for animals applies equally for humans, and have some challenging and thought-provoking indications for humans. It's very comfortable for biologists like Stephen Jay Gould, or Steven Rose, to write about evolution in general and animals in general, but to draw a line around the human mind and try to keep it immune to analysis, try to keep it essentially outside the domain of science itself.

I'm a believer in the unity of science; I don't believe there should be any artificial boundaries drawn around anything. I'm interested in pushing evolutionary theory absolutely as far as it can go into the deepest recesses of the mind, into consciousness, and intimacy and romance, and our self-concept, and things that really matter to us. I'm also interested in pushing it into domains like intelligence that might be politically explosive but are extremely socially important.

It's time we grow up; it's time to face the music and to confront these issues. There's never been a time before when as many people are reading popular science, or watching science television, or expressing an interest in science, and when the sophistication of public understanding is really taking off like now. People are ready to confront these issues, and it's patronizing for some biologist with a vested interest in intellectual status quo to try to keep the human mind out of bounds, to try to keep it outside science.

Science is interesting—it's powerful at what it does, but people credit it with far too much ideological importance. Basically people believe what they want to believe politically. There's even evidence from behavior genetics that mostly people's political ideologies are genetically inheritable. Whatever context you grow up in, to some extent the kinds of attitudes and beliefs you have about political issues and social issues does not seem terribly much affected by the intellectual environment that you're exposed to—people pick up the ideas that fit with their preconceptions and they reject those that don't. It's a great mistake to credit science with too much importance in shaping people's attitudes toward other people, toward government policy, toward social priorities—once you know what social priorities you want to pursue, science is very helpful in suggesting effective ways of pursuing them. But it's a great mistake to confuse science with ideology. Ideologues always pick up whatever science looks like it will fit

Geoffrey Miller

their cause and they distort it and present it and support it and they'll try to use it to convince others, but that doesn't mean that scientists should go around trying to censor themselves for fear that their ideas will be picked up and used by the wrong people. The wrong people always pick up and use any ideas they want in the wrong way. There are so many ideas out there anyway that good people can already do good with the ideas at hand and evil people can do evil with the ideas at hand.

Let's take one rather provocative piece of research. There's some evidence from behavior genetics now, some evidence, not a lot, but a little bit, that happiness itself is somewhat inheritable. If you're extremely reactionary and conservative you could say Ah! See, we can't do anything for people, they'll just be happy or not as they see fit; there's no point in trying to improve people's lives. On the other hand you could be a radical socialist and you could take this as a profound critique of capitalist consumerism—you could say people have been duped into believing that the more stuff they acquire the happier they'll be. That is empirically not the case. You could take it either direction. You could also just say well, pragmatically speaking, if you want happy kids, marry somebody happy. Any different scientific discovery can be taken in a thousand different ideological directions for a thousand different purposes.

5

Rescuing Memory

Steven Rose

Biologist; Emeritus Professor, Department of Life Sciences, The Open University; Author, Lifelines: Biology Beyond Determinism *and* The 21st-Century Brain

ROSE: I started as a schoolkid doing chemistry in the backyard, and went to university to study chemistry, where I discovered that there was this incredibly exciting area called biochemistry, and graduated as a biochemist in Cambridge in the late 1950s. This was a time that Watson and Crick had done their stuff, and Fred Sanger had got the first of his two Nobel Prizes, and the department was awash with champagne. I was pretty arrogant, and so I decided that all the interesting biochemical and genetic questions had been answered, and the next frontier was the brain. So I went off and did graduate study at the Institute of Psychiatry in London and then a postdoc in neuroscience at Oxford, and eventually set up my own laboratory at the Open University, where I've been these past thirty years.

The crucial thing in any sort of science is to find a reliable model that you can study. One of my old advisers, Hans Krebs, the great biochemist, said for any problem in science God has created the right organism to study it. For me it turned out that the right organism for studying brain and memory was the chick because day-old chicks have this tremendous capacity to learn very fast about their environment. Back in the 1970s I did a series of experiments on imprinting in chicks along with Patrick Bateson and Gabriel Horn, which set a sort of standard in the field for how you can study these things.

More recently I moved on to a simpler model. When you give a chick a small bright object, like a little bead, to peck, they will peck at it spontaneously; they're discovering their environment, what's good food, what's bad food, and so on. If you make the bead taste bitter, they'll peck it once, they'll really dislike it, shaking their heads and wiping their beaks, and they won't peck at a bead like that afterward. With one trial, ten seconds, this animal has learned something which lasts a good chunk of its lifetime.

So what's going on in the brain under those circumstances? To study that, you need to pull together all the available techniques to study electrical properties, cellular properties, molecular properties— do the connections in the brain change, can we actually study those by actually looking at them in a microscope? Can we identify the molecules involved in the change, and so on.

Now we're at the point where we know pretty well the molecular cascade of processes that go on when the chick learns this task; what it comes down to in the end is that the chick makes a set of new proteins which stick together the connections between the cells, the synapses, in some new configuration. We've identified this class of proteins, we're analyzing their structures in a variety of ways. It has been an intellectual pursuit of mine for thirty years and it's greatly fascinating.

I always argued that this was a pursuit which was about pure science, and there wasn't going to be a payoff, as it were, except that we would learn more about ourselves and how we work. But what's happened in the last three or four years is something really interesting, because it turns out that among the molecules which were involved in this key process of putting memories together are the substances which get disordered in Alzheimer's disease.

So for the last two or three years a good chunk of the work in our lab has been to look at these molecules to try to understand what they're doing in the brain in terms of how they help stick cells together when memories are being made, how they go wrong when

Steven Rose

the bits of the external sections of the molecules get broken off, and, very excitingly, what you can do to reverse that, or prevent that from happening. One of the things that I was doing at Cold Spring was talking about a new molecule that we've discovered—a little peptide, five amino acids long, which seems to be able to rescue the memory loss that you get with the disorder of the Alzheimer proteins. What started as a sheer intellectual excitement also looks like it's going to have rather significant human payoff, and that's good news.

EDGE: How did you make this discovery?

ROSE: It came about in a rather classical scientific way. What we did first was to show that in order to make long-term memory—that is, memory that persists for more than half an hour or so—you need to make a new class of proteins. Then, using standard biochemical techniques, we were able to identify what the class of proteins were. They turn out to be a group called cell adhesion molecules. That is, they're molecules whose job is to stick together the two sides of the synaptic junction, the business end of the relationship between one cell and another. And that was interesting in itself; you can discover how they work, you can show how you have to unstick them and re-stick them in new configurations.

I was looking at this, and then I suddenly realized that one of the key proteins which is a major risk factor for Alzheimer's disease is itself a cell adhesion molecule. The question was, could that be involved in memory as well? And it turns out that the normal functioning of this molecule is necessary for long-term memory to be made; if we stop the molecule from functioning—you put an antibody into the brain which binds to the molecule, or a specific bit of RNA which stops it being synthesized—then the memories can't be made.

Then if you look at the structure of this molecule, the amyloid precursor protein, it turns out that there is a very small section of it which is just a few amino acids long and seems to have some very special properties. It's those properties which you can mimic by making an artificial peptide, and it turns out it will rescue the memory which is lost otherwise. Clearly that's a long way from having a drug which will cure or protect from Alzheimer's disease. But nonetheless, being able to rescue memory in this sense seems to me to be a step which is potentially in the right direction.

EDGE: What steps are necessary to make this available for human use?

ROSE: There's a lot more work to be done in animals first of all, the standard sort of drug development stuff would have to be done, and then you would have to show that you can take it orally without it breaking down—at the moment we have to get it in by injection. Or else you have to find a way of protecting it so it can get into the brain. Then there are various other bits and pieces of peptide controls that we need to do, and so on. You're talking a few years downstream, but you're moving in that direction.

EDGE: What has been the reaction among scientists?

ROSE: People are pretty excited about it. The formal scientific paper is just in press.

EDGE: When will it come out?

ROSE: Within some months from now, I suspect. What we haven't done is patent it.

EDGE: For ideological reasons?

ROSE: I suppose so, yes.

EDGE: Then somebody will come and patent it.

ROSE: No, they can't once it's published. You can only patent it if it's new. Patent law is a bit different in Europe. I have to say that a fair number of my molecular neuroscience colleagues, mainly on this side of the water, have got companies in which they're trying to develop molecules which will do this sort of thing. That's fine, but I would prefer to publish in the scientific press and develop things that way.

EDGE: Did you see the Arnold Schwarzenegger movie *Total Recall*? How long will it be before you begin to implant memories?

Steven Rose

ROSE: That's a different story. People are unclear about what we mean by a memory-enhancing drug. If you go to the smart bars in San Francisco and buy a smart drug, a memory drug (none of which work, I have to say, but they're good marketing for the health food stores and bars), what you don't get is something which will give you someone else's memory, or even bring back memories you lost when you were just a few years old. What they do is they help in the transition from short- to long-term memory.

If you take someone with Alzheimer's disease, then the first problems that people notice are things like you don't know where you've left your car keys, whether you've done your shopping this morning, whether you know the person who's ringing your front doorbell. The early stages of the disease are stages in which you will remember things for a few minutes but then you will forget them. What we need to do is find a way of helping people in the early stages of Alzheimer's disease to remain in the community rather than have to be in care. To do that they need to be able to hold their short-term memory. Most of the so-called smart drugs are looking at doing that.

Further downstream there's the question of why do people get Alzheimer's disease in the first place? Is there something we can do by way of neuro-protection? Is there something you can take like you take vitamin E or half an aspirin, something like that which will build up some protection? Interestingly, the best evidence for neuro-protection comes not out of the lab but out of epidemiology. It turns out that postmenopausal women who are on HRT are much less likely to get Alzheimer's disease than if they're not on HRT, and that has to do with estrogen, although it's probably not estrogen itself in the brain.

What happens is that the sex hormones, the steroids, are converted in the brain into things called neurosteroids, brain steroids. My guess is that if we're going toward neuro-protection there will be an interaction between these peptides I'm looking at, the neurosteroids, and some other growth factors in the brain. So it will be possible to get a cocktail of processes which will be able to provide neuro protection in this sort of way. That will be the long-term aim.

There are a lot of risk factors for Alzheimer's disease. Some of them are genetic, or in other cases there are genes you've got that are risk factors, and they will interact with things in the environment. The proteins that we're looking at are the risk factors for Alzheimer's disease. They are proteins called presenilin, the amyloid precursor protein, and so on. Somehow there's an interaction between whether you've got these proteins, whether you have some problems—for example, if you've had concussion as a kid, you've been involved in a football game and banged your head or had a car accident, or you've had general anesthesia, you are more likely to get Alzheimer's when you're old than if you've had none of those things. So there's a whole lot of environmental risk factors. How they play together no one knows.

EDGE: How does this line of research play into Darwinian ideas?

ROSE: It depends what you mean by Darwinian ideas, which is one of the problems. Darwin's basic idea is very straightforward. What is not controversial is that evolution occurs. What is at issue is the mechanism of evolutionary change. And the Darwinian evolutionary process says something which is also incontrovertible, that like breeds like with variations, that all organisms produce more offspring than can survive into adulthood and reproduce themselves. Those variations which are best able to survive are more likely to survive into adulthood and breed in their turn, so you get evolutionary change like that. No question. That's one of the fundamental mechanisms of evolutionary change.

But if you read Darwin himself, he was very clear that there are others as well. Sexual selection is one, random changes are another. And chance—the issues that Steve Gould calls contingency—becomes very important here as well. Darwinian mechanisms are very good

Steven Rose

for species getting better at what species do, but they're not good at making new species. Darwin himself was very well able to recognize this, which is why the Galápagos became very important. Here were islands very close together populated by species which seemed similar but had particular differences from island to island.

Much later when he was looking at the specimens that he'd got from the different islands, particularly at finches, of which there are generally agreed to be thirteen different species in the different islands, Darwin came to the conclusion that what must have happened is that the original parents of all these finches had come from the mainland, from Ecuador, which is about 400 miles to the east. Once they were on the islands, they bred and they radiated out. In the different islands there were different potential foods available and the finches became more specialized accordingly—some of them are cactus eaters, some of them are insect eaters, some of them are ground finches, there's a woodpecker finch, there's an insect-eating warbler finch, and so on. All of which probably came from the same original stock. This is one way in which new species were produced, by arriving in a virgin territory and then radiating out from there. So all these mechanisms become very important as far as evolution is concerned.

Now we come to the question that you were asking, which is about the evolution of humans, and the relationship between our brains and our brain processes and evolutionary mechanisms. We are evolutionary products. The particular evolutionary line which has led to humans has been one which has achieved species success by the individuals developing bigger and bigger brains. Now brains aren't necessarily the only way to evolutionary success—bacteria (Lynn Margulis would say proctista) outnumber us, and will probably outsurvive us in the world. But once you start on the evolutionary line which leads to brains, once you're an omnivore, you have to hunt your prey, or you have to learn to escape from prey, then there's an evolutionary pressure to get smarter—that's the route that led to humans.

What our evolution has given us is brains which are enormously powerful and adaptive, capable of enabling us to live in the very complicated social circumstances in which we do, and capable of creating our own history and our own technology. There's a lot of debate which you get from the ultra-Darwinians about free will.

Richard Dawkins ends one of his books by talking about the power of humans, that only we can escape the tyranny of our selfish genes. Somehow free will rescues us from the sort of determinism which is given by our genes. I don't look at it like that.

I don't take free will very seriously. I would say something different, and that is that we have to get rid of this whole attempt to create dichotomies between nature and nurture. The real thing about our brain development, our development as organisms, is not a dichotomy between nature and nurture, but a dichotomy between specificity and plasticity, or perhaps between process and outcome. What is required is a developmental system which is partly not modified by the environment and partly capable of responding to the environment.

Why must it not be modified by the environment? To take a very simple example, a newborn baby's eyes are connected via other brain regions to the visual cortex, at the back of the brain. As the child develops, the eyes grow, the different brain regions grow, and the visual cortex grows—but they grow at different rates. What you've got to do is keep an orderly relationship between the inputs from the eyes finally to the inputs to the visual cortex. Otherwise you'd cease to be able to see or make sense of what you saw. And you don't really want that to be too much screwed around by the environment. So you've got to have specific developmental mechanisms which hold that wiring and make sure that the connections are made and broken in an orderly way. That's specificity.

On the other hand, you've also got to have plasticity, the ability to modify your response to the environment by depending on experience. Take the visual system again as an example—how the visual

Steven Rose

system is finally fine-tuned depends very much on the shapes and patterns that you experience as a young and developing child. Equally we have to learn, and learning means that we have to make and break and remold connections in our brains the whole time.

This intense dynamism, which is fundamental to understanding developmental processes, is lost in the argument that the ultra-Darwinians have that there is, if you like, almost a direct line between a gene and a phenotype, unmodifiable by environmental change. The crucial thing we have to understand, or that I want to understand as a scientist, theoretically and experimentally, is the way that this interplay occurs during development. And that's in a sense what memory is a special case of. Pat Bateson discusses this at some length in his new book, *Design for a Life*.

EDGE: Let's talk about your trip to the Galápagos with Pat.

ROSE: The idea actually came originally from my partner, Hilary. She's a sociologist of science. She said to me one day, we've seen where Marx and Freud lived, and developed their ideas; isn't it time we looked at some of the origins of that other extraordinary founder of modern ideas, and that's Darwin. Of course, you can go to Darwin's house just outside London, and that's interesting enough in itself, but the obvious place to go is the Galápagos. We invited a group of ten social scientists and biologists, chartered a boat called the *Beagle III*—*Beagle I* being Darwin's *Beagle*, *Beagle II* being the boat that was owned by the Charles Darwin Research Center in the Galápagos, which sank a few years ago—*Beagle III* is the new one. And we flew to Ecuador—an Anglo-American–Italian party of biologists, and social scientists, including Hilary and myself, Pat Bateson and his daughter Melissa, who works with starlings, and Ruth Hubbard, who's the biochemist and critic of genetic determinism and gene technology from Harvard. Pat Bateson is the colleague whom I worked with for many years on the imprinting of a chick, and he and I have worked together and shared ideas on many things over most of our lives, as it happens. But he knows a tremendous amount, much more than I do, as an ethologist, a student of animal behavior, about life in the wild.

Pat knows a tremendous amount about bird life. And we managed to recruit for ourselves a brilliant young naturalist guide from the Charles Darwin Research Center who traveled with us. The Research Center is on one of the biggest islands in the Galápagos, Santa Cruz. There are only three inhabited islands within the Galápagos. The whole ecology is of course very fragile. When Darwin visited there the animals were incredibly tame. He describes how the birds would come and sit on his hat—a hawk came and sat on the end of his musket. Of course at that stage he wasn't ecologically sensitive in the sense that you would recognize now. He and his colleagues were only too pleased to kill and eat the animals; they did very well out of killing and eating the land tortoises, for example.

The indigenous population is now threatened on some of the islands, particularly by goats, by farming, by dogs, by feral cats, feral pigs, and so on. Last year's El Niño produced a particular disaster on one of the bigger islands, Isabella, which has two fertile areas, separated by a very large lava field. The goats were confined to one of the fertile areas, and the belief was that they couldn't cross the lava field, and on the other side of the fertile area is a volcano which is a feeding ground for the land tortoises. During El Niño, when the drought came, the goats did in fact cross the lava field, and there are now at least 10,000 of them destroying the vegetation on which the land tortoises are dependent. So there's now an attempt to eradicate them. It's a very interesting argument, why should we privilege tortoises over goats? But somehow we do, because the tortoises live only there, and there are lots of goats—but for the individual goat, or the individual tortoise, I feel the argument is a bit more complicated than that.

In order to preserve the ecology you're not allowed to sleep on the islands. What you can do is go on designated tracks on them with a guide. You can go within a meter of the animals, but of course the an-

Steven Rose

imals come much closer to you because they're still pretty unafraid of humans. It's really a paradise for watching—the sea and land iguanas, the sea lions, the little lava lizards—birds. Frigate birds, the males have these huge red pouches that they blow up in order to attract the females, so it's like flying with sort of a great red balloon strapped to your chest. The tropic birds, the waved albatrosses, the blue-footed boobies, the vermillion fly-catchers, and then of course the Darwin finches themselves—which are small, rather unassuming birds, but nonetheless of course intensely interesting.

Our trip was both fascinating from the point of view of ecology—and I haven't said anything about the flora, only the animals—but also because of the group that was there. There was a continuous seminar and debate about the nature of evolutionary processes—arguments about how what Darwin was writing relates to what the ultra-Darwinists are arguing now. The nature of selection, the nature of adaptation, and so on. It was a fascinating experience apart from being very beautiful.

We tended to travel by night, and probably if one went east to west or north to south it would take about twenty hours in a boat to get from the furthermost extremities of the islands at one end to the other. Some of them are very small, only a few hectares across, others are much larger. They're mainly volcanic. Some of them are almost completely naked lava, and have virtually no vegetation, or other life on them at all. The volcanoes are still active on some of them. In geological terms it's very recent; you're talking over 3 million years, something like that, since the islands were formed. And they're at the intersection of a number of sea currents, the Humboldt Current, the Cromwell Current and others, so the sea ecology is also very interesting.

EDGE: You had some of the world's leading biologists on the boat—how did this group see things differently from Darwin?

ROSE: The issue which obsesses Pat and myself—the relationship between genes and development—has been a problem for biology since Darwin's day. The tragedy for biology is that whereas in the beginning of this century developmental biology and genetics were, if you like, part of one science, as Darwinism and Mendelism became merged, so genetics and developmental biology separated. What I mean by that is this: development became the study of similarities. How is it that we all grow up with two arms, five fingers on each hand, more or less between one and a half and two meters high, and so on. Those extraordinary universalisms. So what is the nature of the developmental process which generates this?

Whereas in contrast genetics became a science of differences. That is, what is it that makes me different from you? One person has brown eyes, one person has blue eyes, or whatever. And yet the two questions are two sides of the same coin. But they've become very polarized in the history of the way the sciences have grown up. I suppose that Pat's concern as a developmental biologist is really the nature of the rules of development. If you look at his new book, he has this cooking metaphor which runs through it, as to how you start, if you like, with the raw materials and you transform them in the cooking process and the cooking is development.

But you can't then separate out the ultimate cooked product into X percent of this or Y percent of the other. Ruth Hubbard and I start as biochemists, and I suppose from our point of view the question is to try to understand what genes actually do in the molecular dance with the cells during development. But what we biochemists mean by genes is very different from the way that theoretical biologists, or evolutionary biologists, like Richard Dawkins, or John Maynard Smith, would talk about them.

For them genes are—I think both of them would agree with this, John certainly would—the genes are almost theoretical postulates, that you look at the ways in which animals behave; you have a mathematical model of their interactions, what would produce evolutionary

Steven Rose

success. Then you postulate a gene that does X, or a gene that does Y. It doesn't matter for you in that sense if there is really a gene, a bit of DNA which does X or does Y. I mean it's a bit that you fit into an equation. And if you look at John's evolutionary stable strategies, or the other models, which are tremendously important, that he's made very much of his own—these are mathematical constructs.

Take an example from a different field—there's an interesting discussion between Roger Penrose and Stephen Hawking in which they debate the reality of the physical models that they're working with. Penrose believes in the reality of the constructions he's handling. Hawking says no, so far as he is concerned, what matters is simply whether the equations work. And if you could make a different set of equations and they could work as well then fine—the important thing is the equations. And it doesn't matter if they map onto "reality"—tables, chairs, molecules, or whatever.

And in a sense, part of the problem with the whole fight in evolutionary biology at the moment goes right the way back to this philosophical distinction; but if you're a biochemist, if you're a neuroscientist like myself, and you get your fingers sticky with real bits of DNA, with real cellular interactions—I work with DNA in the lab—then you see it in a very different way: it matters to you that these are real processes and not just theoretical mathematical models that are going on. There is an element—I mean, okay, in all this dispute, which is partly ideological, partly philosophical, and so on, but there is an element of a difference between whether you are a sticky-fingered biologist who actually does research, or whether you sit and make models of things.

EDGE: What did you learn on the islands? What would you tell Darwin?

ROSE: That's an interesting question. What we would talk about with Darwin are some of the things he began to think about toward the end of his life, especially development. He was very interested, for example, in observing the development of his own children. He's got marvelous descriptions of when his children do particular hand actions, mimic faces, and so on and so forth—that whole process of development and learning is something that I think would be interesting to have a very rich discussion with Darwin about.

Darwin was much more open-minded about these processes than the present ultra-Darwinians like Wilson and the evolutionary psychologists. I believe he would have disagreed with what I construe as their genetic determinism.

One of the things that Pat, Melissa, and I were interested in, and which we kept coming back to all the time in the discussions, is if you see an animal behaving in a particular sort of way, or its behavior being modified depending on the environment in which it is in, how do we understand this, whether we understand it as learning, or the expression of innate developmental rules, and also how much is adaptive.

You see a tremendous variety of colors in tropical fish, for example. Do we have to assume that each of those has actually been selected or not? Let me give you the classical example—Pat uses it in his book and also in his chapter for the book Hilary and I are editing: On some of the islands there are flamingos which are a beautiful pink color. Back even before the present fights, back in the beginning of this century, Thayer, an American naturalist and illustrator, suggested that the pink color of the flamingo was an adaptation so they would be less visible to predators against the setting sun.

But in fact the pink color depends on their diet. If they eat a lot of shrimp they go brighter pink, if they don't eat a shrimp diet they go a paler pink. It would be very hard to argue that the pink color was a Darwinian adaptation to protect against predators rather than an epiphenomenon, a consequence of the diet. That's one of the things that is actually an extremely important issue within biology: what's adaptive, what's not adaptive, what's accidental in some sort of ways. And the flamingos make a very good example.

Steven Rose

Then Pat and I were talking about another one—I keep cats, he breeds cats, and apart from birds he knows a great deal about cat genetics and cat breeding. We all know that a cat sits on your lap and purrs. Is that purring adaptation? Why do cats purr?

The answer is, we haven't the slightest idea, though one can think of it as some type of social signaling mechanism. But how could one study that experimentally? You might exploit natural variation in the amount of purring and look for correlations in the response of social companions, or to try rather brutal surgical interventions to de-purr a cat, and you'd have to deafen its littermates and its mother as it was being reared; you'd have to look at what effect that had on its behavior and its social organization. No one wants to do an experiment like that.

But if you ask why cats purr, and you asked ultra-Darwinians, they would probably say there has to be an evolutionarily adaptive explanation for it—an alternative could be what Steve Gould calls an exaptation, something which arose during evolution for other reasons or by accident and then got co-opted for current purposes; the point is that we need to be much more eclectic and much more open to there being multiple explanations of why anything happens in nature. That there are explanations is clear, but they cannot simply be reduced to the working out of the imperative of the selfish genes.

6

How Is Personality Formed?

Frank Sulloway

Visiting Scholar, School of Personality and Social Research; Author, Born to Rebel: Birth Order, Family Dynamics, and Creative Lives

FRANK SULLOWAY: During the last two decades I have experienced a major shift in my career interests. I started out as a historian of science and was primarily interested in historical questions about people's intellectual lives. In trying to understand the sources of creative achievement in science, I gradually became interested in problems of human development and especially in how Darwinian theory can help us to understand the development of personality. I now consider myself a psychologist, in addition to being a historian.

EDGE: How did you make that leap?

SULLOWAY: This leap was determined by the kinds of questions I was asking. I was initially drawn to the problem of why scientists accept new ideas. If you survey the history of science, it is apparent that most individuals who have accepted radical innovations did not do so simply because they knew of some line of evidence that other people were unaware of. Darwin is a good case in point. He came back from the *Beagle* voyage and displayed his famous Galápagos specimens in London. Within six months of his return, most of the top naturalists in Britain had seen Darwin's Galápagos finches and reptiles, and hence the crucial evidence that converted Darwin to evolution (and that we now consider the textbook case of evolution in action). John Gould, who was one of the greatest ornithologists of the 19th century, knew far more about Darwin's Galápagos birds than Darwin did. Gould corrected numerous mistakes that Darwin had made during the *Beagle* voyage, such as thinking that many of the finches from the Galápagos Islands were the forms that they have come to mimic through biological evolution. For example, Darwin had mistaken the warbler finch for a warbler, and he had thought the cactus finch was a member of the *Icteridae*—a completely different family of birds. Gould corrected these errors and also showed Darwin that some of the other birds he had not recognized as finches were part of a single closely related group. Darwin was stunned by this and other crucial information that he received from Gould in March of 1837, and Darwin immediately became an evolutionist. The strange thing is that Gould did not. He remained a creationist even after *The Origin of Species* was published. Hence the man who knew more saw less, and the man who knew less saw more. It struck me that this puzzling episode in intellectual history had something to do with temperament, or character, or personality. It certainly didn't have anything to do with the scientific evidence per se. Darwin, Gould, and many other contemporary naturalists all knew about the same evidence. This leads to the inference that people who make creative leaps in science, and in other fields, do so in part because of their personalities—and more particularly because of their ability to think in new and unconventional ways. In short, I became interested in psychology.

EDGE: Was this a purely intuitive leap of mind?

SULLOWAY: There was certainly a lot of intuition involved in the leap. Fortunately, the intuitive leap was then followed up by hypothesis testing, which is a method that saves us all from becoming either astrologers or psychoanalysts.

EDGE: How did this idea creep into your consciousness?

SULLOWAY: It was partly intuition, and it was partly just hard evidence. In the early 1970s I began reading everything I could find in personality psychology, especially the literature on cognitive style, and I also began doing research in this area. Eventually I stumbled onto the topic of birth order, on which I subsequently spent two decades doing research. Birth order, however, was just the tip of the iceberg in this research project. The minute one begins to deal with the issue of family dynamics, one also encounters other important factors that are causing personality to develop the way it does.

EDGE: What was your background?

SULLOWAY: I was a first-year graduate student when I developed the interests that have marked my work on scientific creativity. I was just beginning to do my preliminary course work for a degree in the history of science. At that time I anticipated writing a doctoral dissertation on Darwin's life. I had done quite a bit of research on Darwin. For example, I had retraced the *Beagle* voyage around South America and I had made a series of films on Darwin's voyage. I also knew a great deal about Darwin's conversion to evolution, and the specific reasons why Darwin converted; and I had begun to write various papers on these topics—papers that eventually became published articles. In hindsight, I had stumbled onto a problem—Darwin's conversion—that completely changed my career. At one point I seriously considered getting a joint degree in psychology, and did most of the necessary course work in this field. Although I did not end up taking a joint degree, I had entered into what became a kind of hybrid career path. I continued to do considerable reading and research in psychology; I kept up my previous interests in evolutionary biology; and I also continued with my researches in the history of science—particularly on the topic of revolutions in science.

EDGE: Where were you at the time?

SULLOWAY: I was a graduate student at Harvard University. About two years into my graduate studies period I became a Junior Fellow in the Society of Fellows, and this was a wonderful experience. Being a Junior Fellow freed me to work in any area that I wanted. I was no longer under the direct supervision of anyone in my department. It was a terrific experience, and I thrived on the independence it provided.

EDGE: Let's talk about the thesis that led you to your book *Born to Rebel*.

SULLOWAY: Essentially what I stumbled on in 1970, and then empirically verified over a twenty-year period, is that aspects of personality that are under environmental control are strongly influenced by family niches. Birth order is particularly important in this regard, because it is a systematic source of differences in family environments. But birth order is not a *cause*, in and of itself. Rather, it's a surrogate, or a proxy, for patterns of family dynamics that are actually molding personality. For example, firstborns are bigger than their younger siblings. They also are older and tend to have more status. In competition with their siblings, there are certain strategies that eldest children can employ that younger children cannot. A younger child can decide to hit an elder sibling, but this is usually not a smart idea because the elder sibling can hit back harder. In general, firstborns tend to be more aggressive; they use strategies and tactics that take advantage of their greater physical size.

There is an important dimension of personality called "agreeableness/antagonism"—one of the Big Five—that exhibits significant differences by birth order. This birth-order difference reflects the difference in the niches that firstborns and younger children typically occupy. Firstborns tend to occupy the niche of a surrogate parent. Acting as a surrogate parent—that is, assisting with child-rearing duties—is a great way to curry favor with parents. For this reason, firstborns tend to identify more closely with their parents, and they also tend to identify with whatever their parents value. Parents value a child's doing well in school, so firstborns are conscientious, do their homework, generally do better at school, and tend to be overrepresented as academics and in *Who's Who*. The niche of the responsible achiever is particularly likely to be open for an eldest child. Once this niche is taken, it is difficult for a younger sibling to compete effectively for the same niche, although they often try.

Frank Sulloway

The typical strategy of younger siblings is to see whether they can compete successfully in a niche already occupied by an elder sibling. If they cannot, then the best strategy is for the younger sibling to branch out—to become more open to experience—and to try to find some alternative niche where they will not be directly compared with their elder siblings. If an elder brother is a great spear-thrower and a younger cannot top that, they might as well take up the bow and arrow. And if there is another older sibling already specializing in the bow and arrow, then it pays to invent the crossbow. The general rule, then, is do something different that adds value to the family unit as a whole. Like Darwin's famous finches, younger siblings are busy diversifying: They are trying to radiate adaptively away from whatever specialized abilities are already represented by siblings who are older than themselves.

These "contrast effects" between siblings explain the relationship between birth order and certain kinds of creativity. Younger siblings are much more likely to accept radical innovations in science and in social thought. Within their own families, they are at the bottom of the pecking order, so they tend to identify more with the underdog and to champion egalitarian causes. Younger siblings were the earliest backers of the Protestant Reformation, and after it the Enlightenment. Most lost causes in history have been supported by younger siblings and opposed by firstborns. This historical difference goes directly back to the kind of psychological differences in strategic niches that siblings occupy within the family constellation.

EDGE: You have stated that younger siblings have more in common with their peers than with their siblings.

SULLOWAY: On average, firstborns are more similar in personality to firstborns in other families than they are to their own younger siblings. Similarly, a youngest child in one family is often more similar to a youngest child in another family than to his or her own elder siblings. Still, all laterborns are more similar to one another, on average, than they are to firstborns.

EDGE: How did you test this hypothesis?

SULLOWAY: There are several ways of testing it. In my book *Born to Rebel*, I engaged in two major empirical assaults on this problem. The first method of attack involved historical evidence. I gathered data on more than 6,500 participants in major revolutions in science, politics, and social thought. In addition, I arranged for each individual's position in each controversy to be validated by half a dozen or more expert historians. Overall, I asked 110 historical experts to examine my lists of participants in revolutions, and to assess whether these lists were representative of participants as a whole. My experts were also asked to nominate missing individuals, and they rated every participant on a scale of acceptance and rejection. Obtaining these expert ratings involved a tremendous amount of work, in part because I did it in person. I flew a quarter of a million miles around the world as I gathered these expert ratings from scholars in England, France, Germany, Italy, and America. My second line of research involved a reassessment of the birth-order literature as a whole. There are more than 2,000 publications on this subject, and what was needed was a meta-analysis to determine whether there are more significant findings than would be expected by chance. In my meta-analysis I tested specific hypotheses about sibling strategies, using the Big Five personality dimensions as my guide. That is, I expected firstborns—relative to laterborns—to be more (1) conscientious, (2) aggressive, (3) conventional, (4) extraverted in the sense of being dominant (laterborns are more extraverted in the sense of being sociable), and (5) emotionally volatile, in the sense of being quicker to anger. All five of these hypotheses were confirmed by my meta-analysis, which involved a statistical survey of 196 birth-order studies controlled for social class and sibship size.

EDGE: What sort of grant support did you have?

Frank Sulloway

SULLOWAY: My collaboration with my 110 expert raters was done when I was a MacArthur Fellow, and this fellowship was an opportune source of support for my project. Being a MacArthur Fellow was a boon to my ability to get on with the massive amounts of empirical research for this project and to overcome one of the most obvious objections to it, namely: If I have selected the historical samples, why should anyone trust my results? It was essential that the classification of my historical participants as supporters or opponents of radical change be done by people other than myself. As a MacArthur Fellow, I spent every penny of my stipend on research and living expenses.

EDGE: What procedures did you use after you gathered the results?

SULLOWAY: After I had assembled my samples for each of the 121 historical events in my study, I coded every individual for up to 256 different background variables. One of the most unusual features about *Born to Rebel* is that it surveys more than a hundred potential causes of radical thinking, and attempts to rank order these influences in terms of overall influence. Is social class a good predictor of radicalism? This variable is in my database, so I can answer this question: Social class is *not* a good predictor. Is age a good predictor? Yes, age is, just as Max Planck and others have thought, although age is not as good a predictor as either social attitudes or birth order. I also tested a special subset of variables—those related to sibling strategies and family dynamics—many of which also turned out to be significant predictors of radicalism. For example, age spacing between siblings is a significant predictor: Large age gaps between brothers and sisters cause the effects of birth order to dissipate. Conflict with parents is also a significant predictor of radicalism, and it is especially important for firstborns. Laterborns do not need to have the Wicked Witch of the West as a mother in order to become radicals: They have their older siblings to induce this behavioral predilection. But firstborns who grow up in happy families typically identify with parents and authority. Significant conflict with a parent tends to undermine this pattern of identification and causes firstborns to identify instead with the underdog. When I tested all of these different variables simultaneously, the single best predictor of radicalism proved to be birth order. But birth order is hardly the *only* significant predictor. The next two predictors in importance are social attitudes and age, followed by parent-offspring conflict.

EDGE: Your sampling of participants in radical revolutions seems to involve highly accomplished people who were successful enough to become historical figures. Would the same results apply if you had included the average person in your samples?

SULLOWAY: There are two ways we can answer this question. The first is to take my sample of 6,500 historical figures and rank them on a scale of eminence. I have done this, using eighteen different eminence measures. There are some people, such as Darwin and Newton, who are particularly eminent. But when we go down the list, in order of eminence, we come to people who are so obscure that even Newton or Darwin scholars have not always heard of them. After we have stratified individuals by eminence, the question we may ask is whether there is any dilution of a general birth-order effect as we go up or down the scale? In other words, are larger effects associated with eminence? As it turns out, the most obscure people in my sample show virtually the same effects for the influence of birth order as do the most eminent people. It is true that I have not included individuals in my study who are so obscure there is no biographical information about them. But by extrapolation, if there are biases in my study owing to the selection of eminent figures, we should be able to detect their extent when the samples have been stratified by eminence.

The second way to tackle this problem is to study ordinary people. Fortunately, this research has already been done. As I have previously mentioned, there are more than 2,000 published studies on birth order. Much of my own contribution in *Born to Rebel* was to try to make sense out of this extensive literature. This literature has been repeatedly criticized because many of the studies are not well designed or controlled for important background variables. The simplest way to solve these problems is to throw away all the studies that are not well designed. If we take the remaining 196 studies that *are* controlled for class and sibship size, we may ask how many significant findings are there in this set of 196 studies. As it turns out, there are 86 significant findings. The key question, then, is how often would this number of significant findings occur by chance? The procedure used to answer this question is called meta-analysis. The answer is that we would expect to get 86 significant results by chance once in a billion times. In fact, the birth-order literature is in surprisingly good shape compared to most other research areas in psychology.

EDGE: Let's talk about the intellectual antecedents.

SULLOWAY: There is a vast literature on birth order and personality, and, of course, on many of the other variables that I studied in *Born to Rebel*, including gender and parent-offspring conflict. Freud, for example, based his theory of personality development on parent-offspring conflict, and most aspects of family dynamics that I studied have also been extensively studied by other people. In my opinion, one of the most useful contributions of *Born to Rebel* was my effort to simultaneously assess many different influences that theorists from Freud to the present have thought were important.

EDGE: Two questions—What about the only child, and what about women? It seems like all the examples I've heard you talking about are males.

SULLOWAY: I included a chapter in my book on women. In this connection I made a special effort to find historical samples where a substantial proportion of women participated in radical events— precisely so I could say something substantive about sex (and sexual differences). In general, women who end up in the history books as supporters of radical causes tend to be an unusual group. To begin with, they are much more liberal than the average man in the population. They are also more likely to have experienced substantial conflict with a parent, and they are far more likely to have been laterborn (and usually lastborn). In other words, the women who made it into the history books are typically the rebels of the family. These are individuals who boldly transgressed into a man's world because they were not willing to sit there and do what women were generally supposed to do prior to the 20th century. Their first "revolution" was getting into my sample. The *historical* revolution they later participated in, and that brought them to my attention, was a second revolution for them. Because I possess a reasonably large proportion of women in certain radical movements in my study—for example, in the Protestant Reformation and in 61 social reform movements that I studied in American history—I can say with confidence that birth-order effects in radical temperament hold for women as well as men.

Only children pose another interesting question. I view only children as the ideal controlled experiment. They are what it is like to have no birth-order effects at all: Only children have no siblings, hence they have no sibling rivalry. Two predictions follow from these circumstances. One is that only children ought to be intermediate on many personality traits. This follows because they are not being pushed by a younger sibling into being particularly conscientious or aggressive; and they are not being pushed by an elder sibling into being particularly daring or unconventional. Hence, only children ought to be somewhere in the behavioral middle. And this is where they turn out to be. Secondly, only children are free to occupy any niche they wish to in childhood—for example, they do not have to worry about who is going to move in to occupy a niche that they vacate. For this reason, they are free to roam around. As a result, they ought to be more variable than average in their personality traits and interests, and they are. Only children are the most unpredictable group. Their behavior is difficult to predict precisely because their childhood options are greater than for people who grow up with siblings.

EDGE: What were some of the reactions to your book?

SULLOWAY: There have been a variety of reactions to the book, some that I anticipated and some that I did not. One of the most surprising reactions involved the accusation that I was a "determinist." This accusation took two forms: one involving determinism in a general sense and the other involving *genetic* determinism. I was puzzled by both forms of this accusation. If one reads my book carefully, it is obvious that sibling strategies are not strictly "determined." Rather, they are *self*-determined. Individuals have considerable choice as to which strategies they adopt in family life. For example, younger siblings are—on average—less aggressive than their elder siblings, but younger siblings always have the *option* of being aggressive. Nothing stops them from punching an older sibling in the nose. But such aggressive acts are generally ill-considered, because older siblings can punch back harder. Younger siblings learn this lesson early on and behave accordingly. Most of the choices that siblings make in the course of human development are voluntary. Hence, these choices are self-determined. It's really a mincing of words to call such actions "determined." We all know that it is unwise to cross the street when a giant Mack truck is likely to run us over. This fact, to which most of us wisely adapt, does not mean that *all* of our actions are predetermined. In short, some things in life are determined, and other things are not; but I hardly see this circumstance as something to get worked up about.

EDGE: You're talking about probabilities, you're not claiming that every firstborn has these characteristics.

SULLOWAY: Right, mine is a "probabilistic" account of behavior, in part because there are so many different variables that influence personality, including gender, parent-offspring conflict, birth order, and lots more that I document in my book. One can legitimately accuse me of being a *multi*-determinist. My book tells a very complex story and, in this story, there is lots of room for individual choices.

The second form of the determinist accusation directed against my book involved attempts to portray me as a genetic determinist. The few reviewers who tried to make this point did not understand the difference between a purely genetic argument and a developmental one. It is true that *Born to Rebel* is very much a Darwinian book, but this is hardly the same as being an argument for genetic determinism. One of the most subtle features of my argument in *Born to Rebel* is that one can propose a Darwinian argument that is highly environmentalist. Normally we don't hear about these kinds of arguments because this aspect of the story of human development is not well understood.

Here's the argument in a nutshell. Based on Darwinian theory, I argue that offspring are predisposed (genetically) to compete for parental investment. The role of the environment inevitably comes in because individuals—based on the contingencies of birth order, gender, and age spacing—tend to occupy different family niches. This part of the argument is not at all based on genetic determinism. There are no genes for being firstborn or genes for being laterborn. Siblings become very different in large part because different family environments—or niches, if you will—lead them to adopt differing strategies in their efforts to get out of childhood alive. Because firstborns are bigger than their younger siblings, it is easier for them to employ aggressive and tough-minded tactics, which then become part of their personality. This part of the theory is very much an *environmental* and *interactionist* argument. My reasoning in *Born to Rebel* is like Pinker's argument in *The Language Instinct*. There's undoubtedly a hardwired capacity for humans to engage in verbal communication, a capacity that other apes do not possess. But the country we grow up in determines which language we learn to speak. In the same way, we are hard-wired in a Darwinian sense to compete with our siblings for parental investment, but the particular aspects of each person's personality are the product of characteristics of the family environment in which one grows up, just as speaking German in one country and French in another country are appropriate linguistic differences produced by the same language instinct. In short, my argument is not just about nature; nor is it just about nurture—it is a combined nature-nurture argument, in which much of the psychological details are clearly on the environmental side.

Most readers of my book correctly understood this point. In an interview with Ted Koppel on *Nightline*, Stephen Jay Gould emphasized this general logic when he said that birth order provides one of the best demonstrations of the power of the environment and is, on this account, a wonderful antidote to the kinds of genetic determinist arguments espoused in *The Bell Curve*. I find it ironic to have been accused of being a genetic determinist by some people, and yet to have been publicly defended against this accusation by one of the leading critics of such views.

EDGE: Interesting that Gould and Pinker, who frequently disagree, appear to support your ideas. What do the adaptationists— John Maynard Smith, George Williams, Richard Dawkins—have to say about your book?

SULLOWAY: I don't know what Maynard Smith or George Williams think. I gave a lecture on my ideas at the Human Behavior and Evolution Society in 1995, where Richard Dawkins was the keynote speaker, and he seems to have been impressed with the argument. He referred to my paper several times in his keynote speech, at the end of the conference.

EDGE: What about Dan Dennett?

SULLOWAY: After the publication of *Born to Rebel*, Dan sent me a cordial letter saying that he had read my book and that, in general, he agreed with my argument. I am not surprised because, for a sophisticated Darwinian such as Dennett, there is not much that is really controversial about the book. It makes good sense that, if offspring are competing for parental investment, they will devise strategies to implement this competition in their favor.

EDGE: Are there any particular people mounting the attack?

SULLOWAY: The critics have not been connected by any single discipline. The most interesting responses to the book are now coming from psychologists who are busy trying to test and replicate some of my findings. This is becoming an interesting source of potential controversies for the following reasons. There are already more than 2,000 studies on birth order, and more than half of those studies show no significant findings. How can this be, if birth order has an important influence on personality? The answer is twofold. The first part of the answer is that self-report data are not all that reliable. If I had been able to ask Robespierre whether he was a mean and vindictive fellow, I don't think he would have replied in the affirmative. If I had been able to ask Darwin's staunch American opponent, Louis Agassiz, whether he considered himself reluctant to accept new ideas, he would rightfully have said, "No, I am very open to new ideas. I was a pioneer in the development of glaciation theory." Agassiz's openness to the theory of the Ice Ages is not inconsistent, however, with his vehement opposition to evolution. Evolution was a *radical* innovation, whereas glaciation theory was a somewhat *conservative* innovation closely allied to catastrophism. Agassiz later used glaciation theory as a conceptual weapon against evolution, claiming that each Ice Age had extinguished life on earth, requiring a new Creation by God to repopulate the planet. When one asks someone a question such as "Are you open to new ideas?" most people interpret the question in ways that fit their own particular values and biases. We are all open to *some* things. What we want to understand is how do birth order and other influences on personality channel our predispositions to be open to experience in specific ways. Personality tests are not particularly good at capturing these context-sensitive effects.

Frank Sulloway

In *Born to Rebel* I was careful to identify the social and intellectual context of each of the innovations I was studying. For each scientific revolution that I studied, I operationalized the social context in terms of how ideologically radical the innovation was, how long the revolution took to be resolved, and various other measures of "radicalism." These markers of controversiality proved to be excellent predictors of the size of birth-order effects. In addition, these contextual markers were also significant predictors of the effectiveness of other explanatory constructs, such as age, parent-offspring conflict, and social attitudes. In my book, I was continually dealing with person-by-situation interaction effects. Psychologists are now trying to replicate my findings without worrying about the context. Another problem with such studies is that self-report data tend to yield fairly small birth-order effects. We know from considerations of statistical power that one needs a sample of between 500 and 1,000 individuals to be reasonably sure that one is not missing a true effect owing to sampling error. The average study in psychology involves about 250 individuals. Psychologists have been designing studies to test my claims, based on samples of 200 to 400 subjects. These studies are generally incapable of answering the question that the investigators are asking, which is a waste of time and effort. Unfortunately, most psychologists—to this day—do not appreciate the issue of statistical power.

I recently designed a study myself to get around these dual problems of statistical power and self-report biases. The sample already includes about 3,500 subjects, and some of the questions I have asked are aimed at tapping objective indicators of behavior. For example, if I ask individuals to tell me how empathetic they are, using a nine-step scale, I know that I am not always going to get a realistic self-appraisal. In addition, most people don't know where they really lie on an objective measure of empathy. They might know that they are higher than the average person, but they do not know whether they are in the 60th percentile or the 70th percentile—we don't go around wearing "empathy badges" that identify us like men and women. And so there's a lot of imprecision in answers to questions of this sort. Small effects, including those for birth order and other aspects of family dynamics, are easily missed. So what I have done in my study is to include a second set of questions, which ask respondents to rate themselves relative to their friends, spouses, and siblings. Consider the approach entailed in a direct sibling comparison. We generally know (or think we know) whether we're higher or lower than a sibling on most personality traits, and so the method of direct sibling comparison serves to anchor each personality scale with a concrete comparison. We might be in error as to where we place ourselves on such scales—in absolute terms—but we are probably close to the truth in assessing the *relative* difference between ourselves and a sibling. When people compare themselves with a sibling, it turns out that the correlations between birth order and personality are at least twice as large compared with when subjects assesses themselves without reference to anyone else.

EDGE: You're talking about statistical results, but a lot of people are reading your book and thinking about it on the personal level.

SULLOWAY: Well, these two ways of viewing the matter are not inconsistent. I employ statistical techniques and large samples just to be sure that I am right about the relationships I am studying. Once a researcher obtains the correct answer by this method, findings can be illustrated by anecdotes, which represent the level of personal truth that lay readers seek in a book such as mine. Anecdotes have a wonderful power to convey emotional truths. But I do not consider anecdotal evidence to be a proof of anything—on this important point I part company with most historians, who actually think they've proven something when they tell a story. A story proves nothing; it just demonstrates that people have been clever enough to find evidence to fit their hypotheses. The approach I took in *Born to Rebel* involved testing my hypotheses using large statistical samples, and then illustrating the various relationships I had documented by telling one or more stories that brought these relationships to life. For example, laterborns are more likely to challenge the status quo, and they are more likely to cause their parents aggravation by doing all sorts of outrageous things. A person who exemplifies this tendency is Voltaire—he got his start as a poet when his family, to amuse themselves, had Voltaire and his elder brother, Armand, engage in poetry contests. The family soon discovered that Voltaire was a terror at satirical poetry—and he was probably aiming many of his scathing ditties at his elder brother, whom he didn't particularly like. The family put an end to these poetry contests. The father subsequently became concerned that his younger son would end up wasting his life in such an unfruitful profession as literature. "You will starve to death," he warned his son. But a poet had been born, and Voltaire became the richest literary figure in all of 18th-century Europe through the sales of his ribald poems, plays, and books. His brother, Armand, by the way, became a religious fanatic. What is Voltaire most famous for? His scathing critiques of the Catholic church!

Here is another story about Voltaire that I cannot resist telling. Voltaire once witnessed his father having a vehement argument with his gardener. Voltaire's father was a stubborn man. He finally dismissed the gardener, saying to him, "I hope you find an employer who is as gracious and kind as I am." Voltaire thought this remark was ridiculous—that his father, one of the most irascible people he knew, would tell the employee he had just fired that he would be lucky to find another employer as even-tempered as himself. Soon after, Voltaire went to see a play. It turned out that there was a scene in the play just like the one Voltaire had witnessed between his father and the gardener. After the play was over, Voltaire went to see the playwright and asked him if he would substitute, in the next performance of the play, a few words that were closer to his father's own remarks. Voltaire then went home and invited his father to attend the play. His father accepted, and as the father sat through the play, there finally came the scene with the gardener. Voltaire wrote of this episode that "My good father was rather mortified." This story reflects the use of the satirical knife blade, and the turning it in his victim, that Voltaire did to his enemies throughout his career. Some noblemen became so outraged by Voltaire's satirical broadsides that they had him beaten, or arranged for him to have a nice long stay in the Bastille. In any event, these are the kinds of biographical stories that bring a figure like Voltaire alive; and they also illustrate the kinds of unconventional and irreverent qualities that younger siblings have displayed throughout history.

EDGE: How has your own birth order affected your personality and your life?

SULLOWAY: I was the third of four boys, but I'm a functional youngest child because my brother Brook is nine years younger than I am (and from a second marriage). For nine years, I therefore grew up without a younger sibling, and I do not think that Brook had much of an influence on my personality. But my two older brothers did have an influence on me; we were each about two and a half years apart, and there was a lot of fighting among us. I think I have a pretty typical laterborn set of personality characteristics. As someone who has existed as an academic for more than two decades without ever holding a formal job, I have had an unconventional career.

EDGE: Are you familiar with Judith Harris's work on nurture?

SULLOWAY: Yes, she has focused on the influence that peer groups have on children. In response to the findings by behavioral geneticists that most environmental influences are not shared by family members, she and a few other psychologists have argued that the family has only limited influence on personality. An alternative viewpoint, to which I subscribe, is that families do not represent a shared environment. Hence, they influence siblings in different ways, which is not the same thing as having *no* influence. I believe that Harris is correct to emphasize the importance of peer groups, but she is too single-minded when she denies the importance of systematic within-family differences. Actually, the two approaches (family niche theory and peer group influences) overlap in important ways. For example, some family members are probably influenced by their peer groups more than others, and we would especially expect this to be the case for younger siblings because they are more open to experience. It appears that middle children, in particular, are the most closely identified with peer groups rather than with the family. One can perform a very simple test of this claim, as Catherine Salmon did in a recent doctoral dissertation at McMaster University. One asks people to respond ten times to the question "Who am I?" Middle children are significantly less likely than firstborns or lastborns to answer "I am a Brockman" or "I am a Sulloway"—that is, middle children do not identify themselves by using the family label. Why is this? From a Darwinian point of view, we know that middle children are at a disadvantage—they don't have the benefit of being first, which leads to greater parental investment because firstborns are closer to the age of reproduction. The lastborn has the benefit of being the last child the parents are going to have, so parents will tend to invest heavily in this child so that it will not die in childhood. The offspring who tend to get lost in the shuffle are middle children. How do they respond? They become peer oriented. If a person is not favored within the family, it is a wise strategy to build one's bridges to other sources of support.

EDGE: What conclusions will a father or mother take away from your book with regard to the raising of their children?

SULLOWAY: I do not directly address the issue of child rearing in my book, although any reader can draw numerous relevant conclusions on this subject. This is an issue, however, that I do discuss in public lectures. One obvious implication of my researches is that sibling rivalry is not pathological. Many people feel that if rivalry exists among offspring, the parents must have done something wrong. This is mistaken: sibling rivalry predates the dinosaurs. Sibling competition shapes creative behavior—it's part of the process by which children sharpen their endearing little claws and get ready for life. It is a considerable relief for parents to understand this point. Second, parents need to understand why siblings engage in rivalry—such competition is part of the effort to feel special within the family, to feel that one is not discriminated against. Ultimately, sibling competition is all about optimizing parental investment. What each sibling wants is special time with each parent, and when parents provide such moments, it makes children happy. In fact, this is a useful bit of practical information, if parents have not already discovered it. By being different, each sibling is trying to develop a special set of interests, a special niche, causing parents to pay attention to them and to them alone.

EDGE: Where are you headed in your future research?

SULLOWAY: I consider the findings in *Born to Rebel* to be just a preliminary outline of the many problems that we are now facing trying to understand personality development. Also, the book provides only a bare introduction to understanding how we can apply Darwinian theory to understanding all of the learned adaptations of childhood. Adaptations in childhood are not just random; they occur for a purpose, and this purpose is to get one's genes into the next generation. There is a whole class of potential future studies that can be done on these issues. These studies are going to require an even stronger interface between evolutionary biology and developmental psychology. I believe this area of research is going to be a very exciting one for the future.

My own future research is going to be more psychological than historical, so that I can answer some of the questions that I could not answer using historical data. In *Born to Rebel* I developed statistical models that combined the predictive power of birth order, parent-

Frank Sulloway

offspring conflict, temperament, and other variables in explaining what historical figures actually did during times of radical social and intellectual change. We can do a far better job in this regard by working with living individuals because we can ask specific questions about developmental history—for example, the nature of strategies employed in dealing with siblings, and to what extent these strategies (and associated personality characteristics) predict adult behavior. The jump to research on living subjects is a bit like moving from a 19th-century locomotive to a 20th-century jet in terms of the sophistication that one can hopefully achieve, and few of these kinds of studies have been done.

In order to achieve the kind of understanding of families that we need to have, we require studies in which all members of the family are studied simultaneously. When psychologists wanted to study an influence such as birth order in the past, they collected data on first-borns and laterborns selected from different families. We miss too much with this approach. I'll give you an example of why we want to study individuals growing up in the same family. Suppose you are a firstborn. Your usual strategy for dominating your younger siblings would be to act like a tough-minded Clint Eastwood (who, incidentally, is a firstborn, like most of the other Hollywood macho types—John Wayne, Sylvester Stallone, Bruce Willis, and all of the actors who have played James Bond). But suppose a firstborn happens to be shy. Shy people do not generally choose to employ strong-arm tactics—they tend to be retiring and physically timid. And this shy behavioral disposition undermines their ability to occupy the typical firstborn niche. So a shy firstborn is likely to develop a different set of strategies for dealing with siblings. Such individuals might try to keep younger siblings in their place by being moody, or by giving younger siblings who have offended them the cold shoulder. There are many other strategies that people can employ in place of strong-arm tactics. The minute one opts for one set of strategies over another, the door is opened for a younger sibling to adopt some of the strategies that are not being employed. If one is comparing two individuals from different families, one misses these kinds of "coadaptations." It should be kept in mind that personality development takes place on a kind of chess board. The moves that one family member makes are dictated by the moves that have already been made by other family members on the same board. Extraordinary as it may seem, very few studies have been done of personality development from this perspective. From an intuitive psychological point of view—but also from a Darwinian point of view—this is the best way to study human development.

EDGE: Will these studies be conducted in Western countries?

SULLOWAY: Since most psychologists live in the Western world, this is where the bulk of these studies will be done. But since psychologists always love to see cross-cultural replications, we will begin to see studies done in places such as Africa or Southeast Asia. Eventually such studies will be done around the world, and we should definitely expect some interesting twists on the story of human development as we go from one culture to another.

EDGE: Last words?

SULLOWAY: I have to say that I had no idea what I was getting into when I stumbled onto the project that culminated in *Born to Rebel*. Looking back twenty-six years later, it has been one of the most interesting things I possibly could have done. I have never gotten bored trying to understand what makes human beings tick. And to have recognized, two decades into the project, that Darwinian theory was a major player in understanding individual human differences was an exciting insight as well. The mysteries of human development have been a wonderful subject to devote my life to, and I hope to stay interested in these problems, and to continue to make progress in trying to resolve them.

7

Mirror Neurons and Imitation Learning as the Driving Force Behind "the Great Leap Forward" in Human Evolution

V. S. Ramachandran

Neuroscientist; Director, Center for Brain and Cognition, University of California, San Diego; Author, Phantoms in the Brain *and* The Tell-Tale Brain

The discovery of mirror neurons in the frontal lobes of monkeys, and their potential relevance to human brain evolution—which I speculate on in this essay—is the single most important "unreported" (or, at least, unpublicized) story of the decade. I predict that mirror neurons will do for psychology what DNA did for biology: They will provide a unifying framework and help explain a host of mental abilities that have hitherto remained mysterious and inaccessible to experiments.

There are many puzzling questions about the evolution of the human mind and brain:

1. The hominid brain reached almost its present size—and perhaps even its present intellectual capacity—about 250,000 years ago. Yet many of the attributes we regard as uniquely human appeared only much later. Why? What was the brain doing during the long "incubation period"? Why did it have all this latent potential for tool use, fire, art, music, and perhaps even language—that blossomed only considerably later? How did these latent abilities emerge, given that natural selection can only

select expressed abilities, not latent ones? I shall call this "Wallace's problem," after the Victorian naturalist Alfred Russel Wallace, who first proposed it.

2. Crude "Oldawan" tools—made by just a few blows to a core stone to create an irregular edge—emerged 2.4 million years ago and were probably made by *Homo habilis*, whose brain size was halfway (700cc) between modern humans (1,300cc) and chimps (400cc). After another million years of evolutionary stasis, aesthetically pleasing "symmetrical" tools began to appear, associated with a standardization of production technique and artifact form. These required switching from a hard hammer to a soft (wooden?) hammer while the tool was being made, in order to ensure a smooth rather than jagged, irregular edge. And last, the invention of stereotyped "assembly line" tools (sophisticated symmetrical bifacial tools) that were hafted to a handle took place only 200,000 years ago. Why was the evolution of the human mind "punctuated" by these relatively sudden upheavals of technological change?

3. Why the sudden explosion (often called "the great leap") in technological sophistication, widespread cave art, clothes, stereotyped dwellings, etc., around 40,000 years ago, even though the brain had achieved its present "modern" size almost a million years earlier?

4. Did language appear completely out of the blue, as suggested by Chomsky? Or did it evolve from a more primitive gestural language that was already in place?

5. Humans are often called the "Machiavellian primate," referring to our ability to "read minds" in order to predict other people's behavior and outsmart them. Why are apes and humans so good at reading other individu-

als' intentions? Do higher primates have a specialized brain center or module for generating a "theory of other minds," as proposed by Nick Humphrey and Simon Baron-Cohen? If so, where is this circuit and how and when did it evolve?

The solution to many of these riddles comes from an unlikely source . . . the study of single neurons in the brains of monkeys. I suggest that the questions become less puzzling when you consider Giacomo Rizzolatti's recent discovery of "mirror neurons" in the ventral premotor area of monkeys. This cluster of neurons, I argue, holds the key to understanding many enigmatic aspects of human evolution. Rizzolatti and Michael Arbib have already pointed out the relevance of their discovery to *language* evolution. But I believe the significance of their findings for understanding other equally important aspects of human evolution has been largely overlooked. This, in my view, is the most important unreported "story" in the last decade.

The Emergence of Language

Unlike many other human traits, such as humor, art, dancing, or music, the survival value of language is obvious—it helps us communicate our thoughts and intentions. But the question of how such an extraordinary ability might have actually evolved has puzzled biologists, psychologists, and philosophers at least since the time of Charles Darwin. The problem is that the human vocal apparatus is vastly more sophisticated than that of any ape, but without the correspondingly sophisticated language areas in the brain the vocal equipment alone would be useless. So how did these two mechanisms with so many sophisticated interlocking parts evolve in tandem? Following Darwin's lead I suggest that our vocal equipment and our remarkable ability to modulate voice evolved mainly for producing emotional calls and musical sounds during courtship ("croonin a toon"). Once that evolved then the brain—especially the left hemisphere—could evolve language.

But a bigger puzzle remains. Is language mediated by a sophisticated and highly specialized "language organ" that is unique to humans and emerged completely out of the blue, as suggested by Chomsky? Or was there a more primitive gestural communication system already in place that provided a scaffolding for the emergence of vocal language?

Rizzolatti's discovery can help us solve this age-old puzzle. He recorded from the ventral premotor area of the frontal lobes of monkeys and found that certain cells will fire when a monkey performs a single, highly specific action with its hand: pulling, pushing, tugging, grasping, picking up and putting a peanut in the mouth, etc.; different neurons fire in response to different actions. One might be tempted to think that these are motor "command" neurons, making muscles do certain things; however, the astonishing truth is that any given mirror neuron will also fire when the monkey in question observes another monkey (or even the experimenter) performing the same action, for example, tasting a peanut! With knowledge of these neurons, you have the basis for understanding a host of very enigmatic aspects of the human mind: "mind-reading" empathy, imitation learning, and even the evolution of language. Anytime you watch someone else doing something (or even starting to do something), the corresponding mirror neuron might fire in your brain, thereby allowing you to "read" and understand another's intentions, and thus to develop a sophisticated "theory of other minds." (I suggest, also, that a loss of these mirror neurons may explain autism—a cruel disease that afflicts children. Without these neurons the child can no longer understand or empathize with other people emotionally and therefore completely withdraws from the world socially.)

Mirror neurons can also enable you to imitate the movements

of others, thereby setting the stage for the complex Lamarckian or cultural inheritance that characterizes our species and liberates us from the constraints of a purely gene-based evolution. Moreover, as Rizzolatti has noted, these neurons may also enable you to mime—and possibly understand—the lip and tongue movements of others, which, in turn, could provide the opportunity for language to evolve. (This is why, when you stick your tongue out at a newborn baby, it will reciprocate! How ironic and poignant that this little gesture encapsulates half a million years of primate brain evolution.) Once you have these two abilities in place—the ability to read someone's intentions and the ability to mime their vocalizations—then you have set in motion the evolution of language. You need no longer speak of a unique language organ and the problem doesn't seem quite so mysterious anymore.

(Another important piece of the puzzle is Rizzolatti's observation that the ventral premotor area may be a homologue of "Broca's area"—a brain center associated with the expressive and syntactic aspects of language in humans.)

These arguments do not in any way negate the idea that there are specialized brain areas for language in humans. We are dealing, here, with the question of how such areas may have *evolved*, not whether they exist or not.

Mirror neurons were discovered in monkeys, but how do we know they exist in the human brain? To find out we studied patients with a strange disorder called anosognosia. Most patients with a right-hemisphere stroke have complete paralysis of the left side of their body and will complain about it, as expected. But about 5 percent of them will vehemently deny their paralysis even though they are mentally otherwise lucid and intelligent. This is the so-called denial syndrome, or anosognosia. To our amazement, we found that some of these patients not only denied their own paralysis, but also denied the paralysis of another patient whose inability to move his arm was clearly visible to them and to others. Denying one's own paralysis is odd enough, but why would a patient deny another patient's paralysis? We suggest that this bizarre observation is best understood in terms of damage to Rizzolatti's mirror neurons. It's as if anytime you want to make a judgment about someone else's movements you have to run a VR (virtual reality) simulation of the corresponding movements in your own brain, and without mirror neurons you cannot do this.

The second piece of evidence comes from studying brain waves (EEG) in humans. When people move their hands a brain wave called the MU wave gets blocked and disappears completely. Eric Altschuler, Jamie Pineda, and I suggested at the Society for Neurosciences in 1998 that this suppression was caused by Rizzolatti's mirror neuron system. Consistent with this theory we found that such a suppression also occurs when a person watches someone else moving his hand, but not if he watches a similar movement by an inanimate object. (We predict that children with autism should show suppression if they move their own hands but not if they watch someone else. Our lab now has preliminary hints from one highly functioning autistic child that this might be true [*Social Neuroscience* abstracts, 2000].)

The Big Bang of Human Evolution

The hominid brain grew at an accelerating pace until it reached its present size of 1,500cc about 200,000 years ago. Yet uniquely human abilities, such as the invention of highly sophisticated "standardized" multipart tools, tailored clothes, art, religious belief, and perhaps even language, are thought to have emerged quite rapidly around 40,000 years ago—a sudden explosion of human mental abilities and culture that is sometimes called "the big bang." If the brain reached its full human potential—or at least size—200,000 years ago, why did it remain idle for 150,000 years? Most scholars are convinced that the big bang occurred because of some unknown genetic change in brain structure. For instance, the archaeologist Steve Mithen has just written a book in which he claims that before the big bang there were three different brain modules in the human brain that were specialized for "social or Machiavellian intelligence," for "mechanical intelligence" or tool use, and for "natural history" (a propensity to classify). These three modules remained isolated from each other, but around 50,000 years ago some genetic change in the brain suddenly allowed them to communicate with each other, resulting in the enormous flexibility and versatility of human consciousness.

I disagree with Mithen's ingenious suggestion and offer a very different solution to the problem. (This is not incompatible with Mithen's view but it's a different idea.) I suggest that the so-called big bang occurred because certain critical environmental triggers acted on a brain that had already become big for some *other* reason and was therefore "preadapted" for those cultural innovations that make us uniquely human (one of the key preadaptations being mirror neurons). Inventions like tool use, art, math, and even aspects of language may have been invented "accidentally" in one place and then spread very quickly given the human brain's amazing capacity for imitation learning and mind reading using mirror neurons. Perhaps *any* major "innovation" happens because of a fortuitous coincidence of environmental circumstances—usually at a single place and time. But given our species's remarkable propensity for miming, such an invention would tend to spread very quickly through the population—once it emerged.

Mirror neurons obviously cannot be the *only* answer to all these riddles of evolution. After all, rhesus monkeys and apes have them, yet they lack the cultural sophistication of humans (although it has recently been shown that chimps at least *do* have the rudiments of culture, even in the wild). I would argue, though, that mirror neurons are *necessary* but not sufficient: their emergence and further development in hominids was a decisive step. The reason is that once you have a certain minimum amount of "imitation learning" and "culture" in place, this culture can, in turn, exert the selection pressure for developing those additional mental traits that make us human. And once this starts happening, you have set in motion the autocatalytic process that culminated in modern human consciousness.

A second problem with my suggestion is that it doesn't explain why the many human innovations that constitute the big bang occurred during a relatively short period. If it's simply a matter of chance discoveries spreading rapidly, why would all of them have occurred at the same time? There are three answers to this objection. First, the evidence that it all took place at the same time is tenuous. The invention of music, shelters, hafted tools, tailored clothing, writing, speech, etc., may have been spread out between 100K and 5k years ago, and the so-called great leap may be a sampling artifact of archaeological excavation. Second, any given innovation (for example, speech or writing or tools) may have served as a catalyst for the others and may have therefore accelerated the pace of culture as a whole. And third, there may indeed have been a genetic change, but it may not have been an increase in the ability to innovate (nor a breakdown of barriers between modules, as suggested by Mithen), but an increase in the sophistication of the mirror neuron system, and therefore in "learnability." The resulting increase in ability to imitate and learn (and teach) would then explain the explosion of cultural change that we call "the great leap forward" or the "big bang" in human evolution. This argument implies that the whole "nature-nurture debate" is largely meaningless as far as humans are concerned. Without the genetically specified learnability that characterizes the human brain, *Homo sapiens* wouldn't deserve the title "*sapiens*" (wise), but without being immersed in a culture that can take advantage of this learnability, the title would be equally inappropriate. In this sense human culture and human brain have coevolved into obligatory mutual parasites—without either the result would not be a human being (no more than you can have a cell without its parasitic mitochondria).

The Second Big Bang

My suggestion that these neurons provided the initial impetus for "runaway" brain/culture coevolution in humans isn't quite as bizarre as it sounds. Imagine a Martian anthropologist was studying human evolution a million years from now. He would be puzzled (like Wallace was) by the relatively sudden emergence of certain mental traits like sophisticated tool use, the use of fire, art, and "culture" and would try to correlate them (as many anthropologists now do) with purported changes in brain size and anatomy caused by mutations. But unlike them he would also be puzzled by the enormous upheavals and changes that occurred after (say) the 19th century—what we call the scientific/industrial revolution. This revolution is, in many ways, much more dramatic (for example, the sudden emergence of nuclear power, automobiles, air travel, and space travel) than the "great leap forward" that happened 40,000 years ago!!

He might be tempted to argue that there *must* have been a genetic change and corresponding change in brain anatomy and behavior to account for this *second* leap forward (just as many anthropologists today seek a genetic explanation for the first one). Yet we *know* that the present one occurred exclusively because of fortuitous *environmental* circumstances, because Galileo invented the "experimental method" that, together with royal patronage and the invention of the printing press, kicked off the scientific revolution. His experiments and the earlier invention of a sophisticated new language called mathematics in India in the first millennium AD (based on place value notation, zero, and the decimal system) set the stage for Newtonian mechanics and the calculus, and "the rest is history," as we say.

Now the thing to bear in mind is that none of this *need* have happened. It certainly did not happen because of a genetic change in the human brains during the Renaissance. It happened at least partly because of imitation learning and rapid "cultural" transmission of knowledge. (Indeed, one could almost argue that there was a *greater* behavioral/cognitive difference between pre–18th-century and post–20th-century humans than between *Homo erectus* and archaic *Homo sapiens*. Unless he knew better our Martian ethologist may conclude that there was a bigger genetic difference between the first two groups than the latter two species!)

Based on this analogy, I suggest, further, that even the *first* great leap forward was made possible largely by imitation and emulation. Wallace's question was perfectly sensible; it *is* very puzzling how a set of extraordinary abilities seemed to emerge "out of the blue." But his solution was wrong . . . the apparently sudden emergence of things like art or sophisticated tools was not because of God or "divine intervention." I would argue instead that just as a single invention (or two) by Galileo and Gutenberg quickly spread and transformed the surface of the globe (although there was no preceding *genetic* change), inventions like fire, tailored clothes, "symmetrical tools," art, etc., *may* have fortuitously emerged in a single place and then spread very quickly. Such inventions may have been made by earlier hominids, too (even chimps and orangs are remarkably inventive . . . who knows how inventive *Homo erectus* or Neandertals were), but early hominids simply may not have had an advanced enough mirror neuron system to allow a rapid transmission and dissemination of ideas. So the ideas quickly drop out of the "meme pool." This system of cells, once it became sophisticated enough to be harnessed for "training" in tool use and for reading other hominids' minds, may have played the same pivotal role in the emergence of human consciousness (and replacement of Neandertals by *Homo sapiens*) as the asteroid impact did in the triumph of mammals over reptiles.

So it makes no more sense to ask, "Why did sophisticated tool use and art emerge only 40,000 years ago even though the brain had all the required *latent* ability 100,000 years earlier?"—than to ask, "Why did space travel occur only a few decades ago, even though our brains were preadapted for space travel at least as far back as Cro-Magnons?" The question ignores the important role of contingency or plain old luck in human evolutionary history.

Thus, I regard Rizzolatti's discovery—and my purely speculative conjectures on their key role in our evolution—as the most important unreported story of the last decade.

8

A Self Worth Having

Nicholas Humphrey

Psychologist, London School of Economics; Author, Seeing Red *and* Soul Dust

Forty years ago I wanted to solve the problem of consciousness. It seemed to me it would be a shame to leave it for the next generation to get the prize. Consciousness presents the greatest-ever challenge to science; so great, that unless we find an answer soon, science itself is in danger of being humbled. Consciousness—phenomenal experience— seems in many ways too good to be true. The way we experience the world seems unnecessarily beautiful, unnecessarily rich and strange.

I've had various goes at it: approaching the problem from different angles—through neurophysiology, through animal behavior, through social science, through philosophy of mind. My guess is we'll need all these approaches, and more, before we see what consciousness really is and what it's for.

Recently I've been toying with a rather grand idea about why we may need to have conscious qualia in our lives. My idea is that we need them in order to realize just how important we are. Our experience of being conscious encourages us as nothing else could to take ourselves seriously as *selves*. It dramatically affects our whole attitude to the kind of people that we think we are. We find new value in our lives and, just as important, in the lives of other people.

I've come to this on the back of my earlier ideas about the nature of sensation. Some time ago I proposed a theory of how sensations work and why they have their qualitative properties. I argued that sensations derive their characteristic phenomenology from the fact that they are—in evolutionary origin—a kind of bodily action, involving reaching back to the stimulus at the body surface with an evaluative response. Conscious feeling, I suggested, is a remarkable kind of "intentional doing." Feelings enter consciousness not as events that happen to us but as activities that we ourselves engender and participate in.

When a person smells a rose, for example, he responds to what's happening at his nostrils with a "virtual action pattern": one of a set of action patterns that originated far back in evolutionary history as evaluative responses to various kinds of stimulation at the body surface—wriggles of acceptance or rejection. In modern human beings these responses are still directed to the site of stimulation, and still retain vestiges of their original function and hedonic tone; but today, instead of carrying through into overt behavior, they've become closed off within internal circuits in the brain; in fact, the efferent signals now project only as far as the sensory cortex, where they interact with the incoming signals from the sense organs to create, momentarily, a self-entangling, recursive, loop. My theory was that the person's sensation, the way he represents what's happening to him and how he feels about it, comes through monitoring his own signals for the action pattern—as extended, by this recursion, into the "thick moment" of the conscious present.

Now, I still think this is a pretty good idea. Especially because of its potential to explain the underlying functional architecture—even the neurophysiology—of phenomenal experience: the "what it's like" to live in the subjective present of sensations. The sensory loops I identified could create an "as-if" time dimension, so that every moment of consciousness lasts—paradoxically—longer than it actually lasts in physical time.

But there was a puzzle that I had pushed aside. I'd produced a model of reentrant circuits in the brain which might possibly provide the basis for the phenomenology of consciousness. I'd proposed an

Nicholas Humphrey

evolutionary story about how these circuits originated as a kind of bodily activity. But, if truth be told, I'd done nothing to explain *why* evolution had taken this remarkable course; at least I certainly hadn't explained the crucial final stage when the activity in the sensory circuits became self-resonant.

Let's be clear: This final stage can hardly have been an accident. In fact, it must have required very fine tuning of the circuits to produce just the right degree of feedback—which is to say, to produce just the right degree and quality of temporal thickening of consciousness. But what's the point? Why ever should natural selection have gone to so much trouble to create a thick subjective present? Why don't we let conscious time slip by like physical time does? What can be the biological advantage to us of experiencing our own presence in the world in this magically rich way?

So that's what I'm working on now. And what I'm now thinking—though it certainly needs further work—is basically that the point of there being a phenomenally rich subjective present is that it provides a new domain for selfhood. Gottlob Frege, the great logician of the early 20th century, made the obvious but crucial observation that a first-person *subject* has to be the subject *of* something. In which case we can ask, what kind of something is up to doing the job? What kind of thing is of sufficient metaphysical weight to supply the experiential substrate of a self—or, at any rate, *a self worth having*? And the answer I'd now suggest is: *nothing less than phenomenal experience*—phenomenal experience with its intrinsic depth and richness, with its qualities of seeming to be more than any physical thing could be.

Phenomenal experience, surely, can and does provide the basis for creating a self worth having. And just see what becomes possible—even natural—once this new self is in place! As subjects *of* something so mysterious and strange, we humans gain new confidence and interest in our own survival, a new interest in other people, too. We begin to be interested in the future, in immortality, and in all sorts of issues to do with co-consciousness and how far consciousness extends around us.

This feeds right back to our biological fitness in both obvious and subtle ways. It makes us more lively, more fascinating and more fascinated, more determined to pursue lives wherever they will take us. In short, more like the amazing piece of work that humans are. Lord Byron said that "the great object of life is sensation—to feel that we exist, even though in pain." That's the raw end of it. But, at a more reflective level, what keeps us going, gives us courage, makes us aim high for ourselves and our children is the feeling that as human selves we have something very special to preserve.

None of this would have happened if it weren't for those sensory circuits in the brain developing their special self-resonance—a development that was pushed along by natural selection for metaphysics. As I once put it (imitating a famous passage of Rousseau): "The first animal who, having enclosed a bit of the world's substance within his skin, said 'this is me,' was perhaps the true founder of individualized life. But it was the first animal who, having enclosed *a bit of time* within his brain, said 'this is my present,' who was the true founder of subjective being."

I've had the good fortune to be involved as a researcher in opening three different doors onto the problem of consciousness: through neuropsychology, ethology, and aesthetics.

When I was a PhD student in Cambridge in the 1960s, I was at the right place at the right time to make a wonderful discovery: the phenomenon that later became called "blindsight." There was a monkey in Larry Weiskrantz's lab, called Helen, who had had the primary visual cortex at the back of her brain completely removed in a surgical operation. The operation had been done a couple of years earlier, and during the two years since the monkey had seemed to be almost completely blind.

Nicholas Humphrey

However, there were reasons to think this might not be the whole story. And so, one week when I had time on my hands and the monkey wasn't involved in Weiskrantz's research, I decided to find out more. We were both at loose ends. Over several days I just sat by her cage and played with her. And an extraordinary thing happened. I realized that this blind monkey was interacting with me with her eyes. I would hold up a piece of apple and wave it in front of her, and she would reach out and touch my finger and try to get it from me. Within a few days she was transformed from a monkey sitting around listlessly, gazing blankly into the distance, to a monkey who had suddenly begun to be interested and involved in vision again.

I persuaded Larry to let me go on working with Helen. Over the next seven years I took her with me from Cambridge to Oxford, and then back to Cambridge. And she and I developed a remarkable relationship. I was her tutor and she was my apprentice. I encouraged her and coaxed her, trying in every way to help her to realize that actually she *wasn't* blind. I took her for walks in the fields and woods near the laboratory at Madingley near Cambridge. And slowly but surely I taught her to see again. In the end she could run around the room picking up crumbs off the floor, she could catch a fly as it passed by. If you didn't know this monkey had no visual cortex, you would have assumed she had completely normal vision.

Yet I was pretty sure that actually her vision wasn't normal. I knew her too well; we'd spent hours and hours in this strange interaction, with me wondering what it's like to be her. And, though I found it hard to put my finger on what was wrong, my sense was that she still didn't *really believe* that she could see, that she herself was unaware of her capacity for vision. There were telling hints in her behavior. For example, if she was upset or frightened, she'd stumble about as if she was in the dark again. It was as if she could only see provided she didn't try too hard.

In 1972 I wrote a popular paper for *New Scientist*, and on the front cover of the magazine they put the headline under Helen's portrait, "A blind monkey who sees everything." But that surely wasn't right. Not *everything*. My own title for the paper inside the magazine was "Seeing and Nothingness," and I went on to argue that basically this was a kind of seeing we'd never had any inkling of before. Could it be there was no phenomenal experience, no sensation accompanying it? With a monkey, who couldn't describe her inner world, there seemed no way of being sure.

Then, a couple of years later Weiskrantz, spurred on by what we'd found with Helen, moved the research to a new level by showing that a human patient with extensive damage to the visual cortex was equally capable of recovering some degree of vision. But now, with this human patient, it was possible to have him tell the researchers what it was like for him. And, to everyone's astonishment, it turned out that, yes, this was indeed *unconscious vision*—blindsight. The patient believed he was blind, and reported no sensation, and yet he could still *guess* the position and shape of objects in the blind part of his visual field.

As I say, I was lucky. It was a remarkable break for a young student to have, and helped shape me both scientifically and personally. It was a transforming experience: day by day to watch in Helen the emergence of an "impossible" capacity. It was like being a midwife to a miracle. It made me feel good. But those seven years also left a different sort of mark on me. After such an unusually intimate experience, I no longer wanted to do research that involved brain lesions in monkeys. I still respect and admire those who continue to do this kind of work, but I myself wanted to go in a different direction.

In 1974 I had the chance to go and work with Dian Fossey, studying mountain gorillas in the Virunga Mountains of Rwanda. Dian was working for a Cambridge PhD, under Robert Hinde at Madingley, and I was nominally in a position of some authority over her, because

I was assistant director of the lab at the time. I went to stay in her camp for three months, to help her with her research, to answer some questions, and to give advice if I could (although of course it really wasn't my place to give advice to Dian Fossey).

I was in an unusual position: a lab-based experimental psychologist, now given the chance to observe the behavior of apes in the wild. Those were the days when everything was much more relaxed than it is today, when I could set off alone at dawn to track down a particular group, and spend the day, even sometimes the night, with them.

In the grandeur of the mountains, half accepted into the gorilla family, watching and watched by a dozen black eyes, far from any other person, left with my own thoughts, I began musing about an issue that has fascinated me ever since: What's it like, for a gorilla, to be a gorilla? What does a gorilla know about what it's like to be me? How do we read minds?

When we're engaging with other human beings, we hardly notice the extent to which we are involved in mind reading. We take it for granted. But the issue comes into much sharper focus when you find yourself doing the same with other animals who are similar to humans but perhaps not similar enough. It's a real challenge to know whether you're getting it right.

I was trying to understand what it was like to be a gorilla, living in a gorilla family in the forest. The gorillas were, maybe, trying to understand what it was like to be me. Puzzling about what was going on between us, I began to wonder about the special role of *introspection* and reflexive consciousness.

When we imagine what it's like to be another person, we project feelings, sensations, beliefs, and wishes into their minds. But of course we can only do this because we've experienced these very states of mind ourselves. Then, could this perhaps provide a clue as to why it's so important to us to be able to introspect? Could it be that the biological function of introspection—the reason the capacity evolved—is precisely that, by introducing us to how our own minds work, it helps us to read the minds of other people?

It dawned on me that this could be the answer to much that is special about human evolution. We humans—and to a lesser extent maybe gorillas and chimps, too—have evolved to be "natural psychologists." The most promising but also the most dangerous elements in our environment are other members of our own species. Success for our human ancestors must have depended on being able to get inside the minds of those they lived with, second-guess them, anticipate where they were going, help them if they needed it, challenge them, or manipulate them. To do this they had to develop brains that would deliver a story about what it's like to be another person *from the inside*.

Later, I would call this new organ of reflexive consciousness the "inner eye."

Since student days, I've been interested in aesthetics, in value. In fact, my next laboratory-based project after finishing the research with Helen was to investigate whether monkeys have aesthetic preferences. I had a hunch that somehow the study of value must be relevant to understanding consciousness—though I wasn't then sure how.

Here's something to think about. Suppose you were to be turned into a sensationless "zombie": someone who is in every respect exactly like a normal human being except for not having phenomenal consciousness (and all that follows from it)—someone for whom the subjective present never lights up. Would life be worth living anymore?

Early on in my career I got involved in another remarkable case study that threw unexpected, and tragic, light on just this question. A twenty-seven-year-old woman came to London from abroad in 1972 to have an operation to remove cataracts from her eyes. She'd been blind since the age of three. The doctor who operated on her had promised her that there was a good chance of being able to see normally again. I met her several months after the operation and found her in a state of great despair. She was convinced the operation was a complete failure; she couldn't see any better than she could before.

Unfortunately, it seemed all too likely that, as the result of years of lack of use, her visual cortex had in fact atrophied, so that she was in effect in much the same condition as my monkey, Helen. And yet, if this were the case, perhaps not all was lost. Perhaps she would be capable of learning to see again as Helen had.

I decided to try some of the same things with her. I took her out into St. James's Park and around London. We walked through the gardens while I described the sights and held her hand. And soon enough it became clear that she did indeed have a capacity for vision that she wasn't aware of. She could point to a pigeon on the grass, she could reach for a flower, she could step up when she came to a curb.

It seemed that, after all, the operation had not been a total failure: her eyes were working again, to a degree. But was this what she was hoping for? No, it only proved the more traumatic. For the awful truth, she let on, was that her vision—just as in blindsight (and very likely it was a kind of blindsight)—still lacked any qualitative dimension. She'd been living for twenty years with the idea of how marvelous it would be if only she could see like other people. She had heard so many accounts, stories, poetry, about the wonders of vision. Yet now here she was, with part of her dream come true, and now she simply *couldn't feel it*. She was desperately disappointed, almost suicidal. In the end she dealt bravely with her situation by putting on her dark glasses again, taking up her white cane, and going back to her former status of being blind.

This case stayed with me—to remind me, if I should ever forget, how much consciousness *matters*. Even to the extent that mattering may be one of the main reasons why consciousness exists. What if it's consciousness that gives us a reason for waking up every day, and going out into the world—to experience the qualia of a rainbow, the sunset, music, interactions with our friends, sex, food? What if consciousness provides such an incentive for living that, as human beings, we would not—and probably could not—do without it?

Of course human beings find meaning on lots of other levels. But the more I try to make sense of it, the more I come back to the fact that we've evolved to regard consciousness as a wonderfully good thing in its own right—which could just be because consciousness *is* a wonderfully good thing in its own right!

You asked me to explain how I've developed as a scientist. I'll confess I've long had the ambition to make a difference—so that I leave the world a different place than it would have been without me. But it's a funny thing: this kind of personal ambition can actually lead to an anxiety about being a scientist. The problem is that scientific truth is thoroughly impersonal. The answers to the questions scientists ask are in a sense already out there in the book of nature, waiting for someone—anyone—to find the key to reading them. If one particular scientist doesn't find the answer today, we can be pretty sure that another one will tomorrow or the next day—probably rather quickly given the way in which science is moving. So, if we're honest, we have to admit that even though we may have great fun in getting to the answer, and maybe great success and fame at having been the first to get there, in the end our personal contribution hardly matters. Even worse, perhaps what we are doing is to make some other scientist miserable—because we beat him to it.

This surely makes science a rather different enterprise from other great enterprises of our culture. Consider painting, or writing, or making music, where it's certainly arguable that every creation is an individual work, which has the stamp of the person and the personality that made it. If Shakespeare hadn't written *Hamlet*, nobody would have written *Hamlet*. If Picasso hadn't painted the *Demoiselles d'Avignon*, nobody would have painted it. If Metallica hadn't composed their heavy metal music, nobody would have done it.

Nicholas Humphrey

Nonetheless, science does at least have this advantage over the arts: we need not doubt that the questions we are asking are important. Everybody wants to know the answer to the riddle of consciousness, or the origin of the universe. By contrast not everybody wants to know the answer to whatever it is that Metallica is trying to do. Or even Picasso. Shakespeare? Well, perhaps Shakespeare's in a different league. Everyone does want to know the answers to the problems posed in *Hamlet*.

The ideal life, maybe, would be to create science, but in a style and with a way of presenting it which does have some of the qualities of great art, but which nonetheless has the security of providing verifiable answers to the big problems. Could someone be the Metallica of science? Maybe that's not the best model. But I'd say someone could certainly be the Dostoyevsky of science.

But, then, does it really matter whether your contribution has your personal stamp on it? I'd be the first to agree there are ways of making a difference—perhaps nobler ways—where who cares whether it's you or someone else. As scientists we have unrivaled opportunities to do things which, through their practical effects, make the world a better place to live in. And in my own work I'm afraid to say this practical element has been very much missing. I've made people interested and excited about ideas, but I can't claim to have done much to change people's lives for the better in any material way.

Maybe it's not too late. Recently I've been involved in research on the placebo effect, coming at it from a mixture of philosophical and evolutionary perspectives. The placebo effect is a very important aspect of all medicine. A large part of medical cures are effected by the patients themselves, when the medical procedure allows the patients to bring their own resources to bear to solve the problems. In the classical placebo case, you give a sugar pill and the patient uses this as an excuse to cure himself. But placebos are actually present in every kind of medical treatment. To the extent the patient believes the treatment is going to work, he allows himself to deploy his own healing resources in a way that he wouldn't have done otherwise.

How should we understand this? What questions should a science of the placebo effect be asking? Of course it's important to investigate the brain mechanisms that underlie these effects, and lots of researchers are already beginning to home in on the problem at the level of neurophysiology and immunology. But it's no less important to look at the bigger picture, and ask: Whatever is going on here, from a *functional* standpoint? If a placebo is releasing in people an ability to cure themselves, why don't they just get on with it? Why ever should anyone withhold self-cure? You'd think that when you're sick you should just get better if you can; you shouldn't need to wait for permission from a doctor, a shaman, or a psychotherapist to utilize your own resources.

It's this level of question that has set me looking for some possible evolutionary explanation. Why should humans and other animals hold healing resources in reserve? What can be the advantages of *not* getting better when you actually could? As I've looked further, I've found many examples of it.

People may die from cancer when they have immune resources still waiting in reserve which could have been deployed against the cancer. People die in head-on car collisions because they don't apply the brakes hard enough. When athletes are running a marathon, they may reach the end of what they can do and collapse from fatigue, when, in fact, their muscles still have significant reserves left in them.

What's going on?

You'll have guessed the way I want to go with this: my idea is that nature has *designed* us to *play safe*, and never to use up everything we've got—because we never know what might still lie around the corner. When we reach the end of a marathon there may still be a lion waiting at the finishing post that's going to suddenly give chase. When we're sick with an infection and respond with an immune re-

Nicholas Humphrey

action, we may still be hit by a further infection the next day. Remember the story of the wise and foolish virgins and their lamps: It's always wise to keep something in reserve.

I'm now thinking in terms of there being what I call a "natural health management system," which does a kind of economic analysis of what the opportunities and the costs of self-cure will be—what resources we've got, how dangerous the situation is right now, and what predictions we can make of what the future holds. It's like a good hospital manager who has to choose if and when to throw resources against this or that problem, to hold so much back, to decide if it's essential to build up this area or that area—basically to try to produce an optimal solution to the problem of maintaining health with enough left over to meet coming challenges.

If this is right, it makes the placebo effect fit into a much larger picture of homeostasis and health management. And it converges with ideas being developed by researchers coming from quite different disciplines. I've been particularly struck by the work of the South African physiologist Timothy Noakes, who has come up with the idea of there being what he calls "a central governor" in the brain which regulates just how far the body should be allowed to go in meeting the demands of extreme exercise.

These ideas are big, because they are producing a new perspective on how we and other animals have evolved to manage our internal healing resources across the board. But it already goes much beyond mere theory.

There's a phenomenon, well known to sports physiologists and athletes, called "interval training." If you want to improve your prowess as an athlete, one highly effective method of doing it is to build up in the following way: If you're a sprinter, for example, you sprint for two minutes and then relax and jog for five minutes. Then you repeat this pattern again, and again. The result is that you soon find you can run about 15 percent better than you could before.

Why does this work? According to Tim Noakes, what may be happening is this: In order to improve peak performance you need to persuade your central governor to let you go beyond your own self-imposed limits, when otherwise "cautionary tiredness" would kick in and say, "No more." And one way of doing this is by teaching your central governor that the risks are not actually so great after all. Through interval training you can teach your own brain that you are not going to get into trouble by pushing yourself a little further than you might otherwise have done.

Noakes's theory is a clever way of looking at how to stretch the limits of athletic performance. But what about applying the same idea in other areas? In particular, what about the possibility that we could have interval training for the immune system? If people are not deploying their immune resources to maximum extent, so that they don't get better when they could have, could we teach them by a similar schedule of exercise for the immune system that it's safe to do so?

Here's the experiment: Let's do it in mice before we try it in humans. We give a mouse a bacterial infection. The mouse gets sick, and throws its immune resources against the infection—but only so far as it dares. Twenty-four hours later we follow up with antibiotics, and the mouse gets better. So the mouse's health management system gets the message that it's safe to go at least this far. Now, a week later, we repeat this pattern of infection followed by relief. Then we do it again, and again. And what I'd hope we'd find is that the mouse's health management system will learn that it can afford to use more of its resources than it otherwise would have dared to, because every time it goes to its own self-imposed limits it discovers it's followed by safe recovery.

Now, suppose we take one mouse which has been put through this regime, and another mouse which hasn't, and we inject them both with a carcinogen. I predict that the mouse which has been through interval training for its immune system will survive the cancer in a way in which a mouse that hasn't done won't.

If this were to work with people, imagine how it might turn medicine around! It might prove to be one of the best ways ever of achieving one of the main goals of modern medicine, which is to get people to use their own healing resources to greater and better effect than they usually do.

I have to say I really like the idea. Maybe this interview will be remembered as its first airing (I hope!).

9

You Can't Be a Sweet Cucumber in a Vinegar Barrel

Philip Zimbardo

Psychologist, Stanford University; Author, The Lucifer Effect: Understanding How Good People Turn Evil

For years I've been interested in a fundamental question concerning what I call the psychology of evil: Why is it that good people do evil deeds? I've been interested in that question since I was a little kid. Growing up in the ghetto in the South Bronx, I had lots of friends who I thought were good kids, but for one reason or another they ended up in serious trouble. They went to jail, they took drugs, or they did terrible things to other people. My whole upbringing was focused on trying to understand what could have made them go wrong.

When you grow up in a privileged environment you want to take credit for the success you see all around, so you become a dispositionalist. You look for character, genes, or family legacy to explain things, because you want to say your father did good things, you did good things, and your kid will do good things. Curiously, if you grow up poor you tend to emphasize external situational factors when trying to understand unusual behavior. When you look around and you see that your father's not working, and you have friends who are selling drugs or their sisters are in prostitution, you don't want to say it's because there's something inside them that makes them do it, because then there's a sense in which it's in your line. Psychologists and social scientists who focus on situations more often than not come from relatively poor, immigrant backgrounds. That's where I came from.

Over the years I've asked that question in more and more refined ways. I began to investigate what specific kinds of situational variables or processes could make someone step across that line between good and evil. We all like to think that the line is impermeable—that people who do terrible things like commit murder, treason, or kidnapping are on the other side of the line—and we could never get over there. We want to believe that we're with the good people. My work began by saying no, that line is permeable. The reason some people are on the good side of the line is that they've never really been tested. They've never really been put in unusual circumstances where they were tempted or seduced across that line. My research over the last thirty years has created situations in the laboratory or in field settings in which we take good, normal, average, healthy people—more often than not healthy college students—and expose them to these kinds of settings.

For example, think of William Golding's *Lord of the Flies*, in which the key variable in transforming Jack Merridew, a good choirboy, into a kid who could not only kill pigs but also then kill Piggy the intellectual is that he changes appearance. He gets naked, uses berries to mask himself, and makes other kids do the same. Then they do something that has been prohibited; namely, they kill pigs they need for food. Once killing is disinhibited, then they are able to kill freely. Is that idea a novelist's conceit, or is it a psychologically valid concept?

To investigate this I created an experiment. We took women students at New York University and made them anonymous. We put them in hoods, put them in the dark, took away their names, gave them numbers, and put them in small groups. And sure enough, within half an hour those sweet women were giving painful electric shocks to other women within an experimental setting. We also repeated that experiment on deindividuation with the Belgian military, and in a variety of formats, with the same outcomes. Any situation that makes you anonymous and gives permission for aggression will bring out the beast in most people. That was the start of my interest in showing how easy it is to get good people to do things they say they would never do.

I also did research on vandalism. When I was a teacher at NYU I noticed that there were hundreds and hundreds of vandalized cars on the streets throughout the city. I lived in Brooklyn and commuted to NYU in the Bronx, and I'd see a car in the street. I'd call the police and say, "You know, there's a car demolished on 167th and Sedgwick Avenue. Was it an accident?" When he told me it was vandals, I said, "Who were the vandals? I'd like to interview them." He told me that they were little, black, or Puerto Rican kids who come out of the sewers, smash everything, paint graffiti on the walls, break windows, and disappear.

So I created what ethologists would call "releaser cues." I bought used cars, took off license plates, and put the hood up, and we photographed what happened. It turns out that it wasn't little, black, Puerto Rican kids, but white, middle-class Americans who happened to be driving by. We had a car near NYU in the Bronx. Within ten minutes the driver of the first car that passed by jacked it up and took a tire. Ten minutes later a little family would come. The father took the radiator, the mother emptied the trunk, and the kid took care of the glove compartment. In forty-eight hours we counted twenty-three destructive contacts with that car. In only one of those were kids involved. We did a comparison in which we set out a car a block from Palo Alto, where Stanford University is. The car was out for a week, and no one touched it until the last day, when it rained and somebody put the hood down. God forbid that the motor should get wet.

This gives you a sense of what a community is. A sense of community means people are as concerned about any property or people on their turf because there's a sense of reciprocal concern. The assumption is that I am concerned because you will be concerned about me and my property. In an anonymous environment nobody knows who

I am and nobody cares, and I don't care to know about anyone else. The environment can convey anonymity externally, or it can be put on like a Ku Klux Klan outfit.

And so I and other colleagues began to do research on dehumanization. What are the ways in which, instead of changing yourself and becoming the aggressor, it becomes easier to be hostile against other people by changing your psychological conception of them? You think of them as worthless animals. That's the killing power of stereotypes.

I put that all together with other research I did thirty years ago during the Stanford prison experiment. The question there was, what happens when you put good people in an evil place? We put good, ordinary college students in a very realistic, prison-like setting in the basement of the Psychology Department at Stanford. We dehumanized the prisoners, gave them numbers, and took away their identity. We also deindividuated the guards, calling them Mr. Correctional Officer, putting them in khaki uniforms, and giving them silver reflecting sunglasses like in the movie *Cool Hand Luke*. Essentially, we translated the anonymity of *Lord of the Flies* into a setting where we could observe exactly what happened from moment to moment.

What's interesting about that experiment is that it is really a study of the competition between institutional power versus the individual will to resist. The companion piece is the study by Stanley Milgram, who was my classmate at James Monroe High School in the Bronx. (Again, it is interesting that we are two situationists who came from the same neighborhood.) His study investigated the power of an individual authority: Some guy in a white lab coat tells you to continue to shock another person even though he's screaming and yelling. That's one way that evil is created, as blind obedience to authority. But more often than not, somebody doesn't have to tell you to do something. You're just in a setting where you look around and everyone else is doing it. Say you're a guard and you don't want to harm the prisoners—because at some level you know they're just college students—but the two other guards on your shift are doing terrible things. They provide social models for you to follow if you are going to be a team player.

In this experiment we selected normal, healthy, good kids that we found through ads in the paper. They were not Stanford students, but kids from all over the country who were in the Bay Area finishing summer school. A hundred kids applied, we interviewed them, and gave them personality tests. We picked the two dozen who were the most normal, most healthy kids. This was 1971, so these were peaceniks, civil rights activists, and antiwar activists. They were hippy kids with long hair. And within a few days, if they were assigned to the guard role, they became abusive, rednecked prison guards.

Every day the level of hostility, abuse, and degradation of the prisoners became worse and worse and worse. Within thirty-six hours the first prisoner had an emotional breakdown, crying, screaming, and thinking irrationally. We had to release him, and each day after that we had to release another prisoner because of extreme stress reactions. The study was supposed to run for two weeks, but I ended it after six days because it was literally out of control. Kids we chose because they were normal and healthy were breaking down. Kids who were pacifists were acting sadistically, taking pleasure in inflicting cruel, evil punishment on prisoners.

That study has legs even today, especially because of the recent exposé of abuses in the Iraqi prison Abu Ghraib. But the study was popular even before then because in a way it's a forerunner of reality TV. You take a bunch of boys, put them in a situation, and videotape them hour after hour. We have visual records of the dramatic transformation of these ordinary kids into brutal, sadistic monsters or pathological zombies—in a DVD format entitled *Quiet Rage: The Stanford Prison Experiment*. The prisoners who remained and did not break down just let the guards do whatever they wanted to them. It's really like a Greek drama more than an experiment, because it's what happens when you put good people in an evil place. Does the place win, or do the people? Answer: Place one, People zero.

We made the study very dramatic. The arrests were made by the city police in their squad cars with sirens wailing. Actual policemen brought the prisoners down to the police station in handcuffs and did the booking. We had visiting days with parents. We had Catholic priests and chaplains. We had public defenders. Although it was an experiment in a basement at Stanford University, we had all of the trappings of a prison. We created a psychologically functional equivalent of the sense of imprisonment. That's why it had this big impact in such a short time.

There are stunning parallels between the Stanford prison experiment and what happened at Abu Ghraib, where some of the visual scenes that we have seen include guards stripping prisoners naked, putting bags over heads, putting them in chains, and having them engage in sexually degrading acts. And in both prisons the worst abuses came on the night shift. Our guards committed very little physical abuse. There was a prisoner riot on the second day, and the guards used physical abuse, and I, as both the superintendent of the prison as well as the principal investigator—My big mistake. You can't play both of those roles; I continually told them that they could not use physical abuse. But then they resorted entirely to psychological controls and psychological domination. There is an interesting comparison between police detectives who, after being forced to give up brutal third-degree abuses in getting confessions, switched to psychological tactics—and they were equally effective in obtaining confessions after interrogation, as my research in the sixties documented.

Our guards would say things to the prisoners like, "You're Frankenstein. You're Mrs. Frankenstein. Walk like Frankenstein. Hug her. Tell her you love her." And then they would push them together.

We learned that in real prisons one of the things guards try to do is

Philip Zimbardo

weaken the masculinity in dominance, because prisoners are a threat to the guards' security. And so at Stanford the prisoners wore smocks with no underpants, like dresses. We did that purposely to feminize them. The guards would tell the prisoners that they should line up to play leapfrog. It's just a simple game, except when you leap over each prisoner your genitals smack each guy's head. Then they'd say, "You, bend over. You're female camels." And when they did their behinds were showing. And then they would tell others, "You're male camels. Line up behind them. Okay, hump them." This is a funny play on words, of course, but they had the prisoners simulating sodomy.

These are exact parallels between what happened in this basement at Stanford thirty years ago and at Abu Ghraib, where you see images of prisoners stripped naked, wearing hoods or masks as guards get them to simulate sodomy. The question is whether what we learned about the psychological mechanisms that transformed our good volunteers into these creatively evil guards can be used to understand the transformation of good American army reservists into the people we see in these "trophy photos" in Abu Ghraib. And my answer is yes, there are very direct parallels.

One of the distressing things I have to think about is whether or not the results of my research, which I've written about extensively, have been incorporated by the Pentagon in its various programs. I hate to think that my research actually contributed to creating this evil, rather than simply helping to explain it. But the situation we have now is that the army, the Pentagon, and the administration are trying to disown any influence on the specific guards seen in those trophy pictures. One of the many investigations (the Schlesinger report) into these abuses explicitly states that the Stanford prison experiment should have served as a forewarning to those running Abu Ghraib Prison of the potential dangers of excesses by guards in such settings.

It is hard to comprehend what the soldiers were thinking when they took photos of themselves engaged in those abuses—trophy photos. I call them trophy photos because the analog is to big game hunters displaying their victory over the beasts of the earth and sea. But a more potent parallel are the trophy photos from lynchings of black men and women over decades. There's a remarkable book called *No Sanctuary*, which shows that for a hundred years not only did Americans lynch blacks in the South and the Midwest, but they often took photos of their illegal lynchings—that often included photos of all those people who were involved. These were not only lynching photos, but brutal whippings, and they also burned blacks alive. In some of the pictures young children are photographed watching the spectacle. To make the horror worse, these images were put on postcards, and people would send them to one another, or would frame them and hang them in their living rooms. Talk about dehumanization! The concept of lynching or burning somebody alive is horrible enough, but then to take a picture, put yourself in it, and then send it to your mother to say, "I'm the third one on the left," is just evil.

These terrible deeds form an interesting analog in America, because there are two things we are curious to understand about Abu Ghraib. First, how did the soldiers get so far out of hand? And second, why would the soldiers take pictures of themselves in positions that make them legally culpable? The ones that are on trial now are the ones in those pictures, although obviously there are many more people involved in various ways. We can understand why they did so not only by applying the basic social-psychological processes from the Stanford prison study, but also by analyzing what was unique in Abu Ghraib.

There are several important concepts. First, in both cases there's the deindividuation, the sense of anonymity. The CIA agents, the civilian interrogators, never wore uniforms or showed identification. In all of the pictures the soldiers were typically not wearing uniforms. They often had their tops off. That's a violation of military protocol, because even in a prison you're supposed to be wearing your uniform. In the 1970s the police would do that during student riots against the Vietnam War. They would take off their jackets with their names and ID numbers. I was at Columbia University in a police riot, and I was at Stanford in a police riot, and the first thing the cops did was to take off anything that identified themselves, or put on gas masks when there was no gassing, only to create a state of anonymity.

At Abu Ghraib you had the social modeling in which somebody takes the lead in doing something. You had the dehumanization, the use of labels of the other as inferior, as worthless. There was a diffusion of responsibility such that nobody was personally accountable. The Stanford prison study identified a whole set of principles, all of which you can see are totally applicable in this setting.

The other thing, of course, is that you had low-level army reservists who had no "mission-specific" training in how to do this difficult, new job. There was little or no supervision of them on the night shift and there was literally no accountability. This went on for months in which the abuses escalated over time. This also happened in my study. Each day it got worse and worse.

And then there is the hidden factor of boredom. One of the main contributors to evil, violence, and hostility in all prisons that we underplay is the boredom factor. In fact, the worst things that happened in our prisons occurred during the night shifts. Guards came on at ten o'clock and had eight hours to kill when nothing was happening. They made things happen by turning the prisoners into their playthings, not out of evil motives, but because this was what was available to break through the boredom. Also at play in the prison in Abu Ghraib was extreme fear among the guards because of the constant mortar attacks that had killed soldiers and prisoners, and escape attempts.

Dehumanization also occurred because the prisoners often had no prison clothes available, or were forced to be naked as a humiliation tactic by the military police and higher-ups. There were too many of them; in a few months the number soared from 400 to over a thousand. They didn't have regular showers, did not speak English, and they stank. Under these conditions it's easy for guards to come to think of the prisoners as animals, and dehumanization processes set in.

When you put that set of horrendous work conditions and external factors together, it creates an evil barrel. You could put virtually anybody in it and you're going to get this kind of evil behavior. The Pentagon and the military say that the Abu Ghraib scandal is the result of a few bad apples in an otherwise good barrel. That's the dispositional analysis. The social psychologist in me, and the consensus among many of my colleagues in experimental social psychology, says that's the wrong analysis. It's not the bad apples, it's the bad barrels that corrupt good people. Understanding the abuses at this Iraqi prison starts with an analysis of both the situational and systematic forces operating on those soldiers working the night shift in that "little shop of horrors."

Coming from New York, I know that if you go by a delicatessen, and you put a sweet cucumber in the vinegar barrel, the cucumber might say, "No, I want to retain my sweetness." But it's hopeless. The barrel will turn the sweet cucumber into a pickle. You can't be a sweet cucumber in a vinegar barrel. My sense is that we have the evil barrel of war, into which we've put this evil barrel of this prison—it turns out actually all of the military prisons have had similar kinds of abuses—and what you get is the corruption of otherwise good people.

I was recently engaged as an expert witness for the defense of one of the Abu Ghraib night shift guards in his court-martial trial. As such, I had access to all of the horror images that various soldiers took of their infamous deeds in action, along with most of the reports of the official investigations; spending a day with the defendant and his wife; arranging to have various psychological assessments made; and checking on his background and army reserve record.

Philip Zimbardo

In addition to realizing the relevance of my earlier research to understanding some of the forces acting on him and the other night shift soldiers, it became apparent that he was also totally abused by the situation that the military had thrust upon him. Imagine the cumulative stress of working twelve-hour night shifts, seven days a week, with not a day off for forty days! Also regularly missing breakfast and lunch because he slept through them, having finished his tour of duty at 4:00 a.m. and sleeping in a small cell in another part of the prison that he rarely left. When he complained about children mixed with adult inmates or mentally ill or those with contagious TB among the prisoners, he was reprimanded, but rewarded for helping to get confessions by softening up the inmates. Not once was there any official supervision on his night shift that he could rely on. There were insufficient guards, 8 for 1,000 inmates, and none had been adequately trained for this tough job. His psychological testing and my interview revealed a young man who had not a single symptom of pathology that he brought into that prison; the situation was the pathological ingredient that infected his reason and judgment. Indeed, in many ways, this soldier is an American icon—good husband, father, worker, patriotic, religious, with many friends and a long history of having lived a most normal, moral small-town life.

Despite my detailed trial testimony about all the situational and systemic forces influencing his distorted group mentality, the judge threw the book at him, giving him eight years in prison and many other penalties. He refused to acknowledge what many of the official investigations clearly revealed, that the abuses at Abu Ghraib could have been prevented or would not have occurred were it not for "a failure or lack of leadership." Some reports list the officers and agencies responsible by name, but they are likely never to be considered bad apples, but only the custodians of a barrel that had some defects. The judge, and juries, in recent military jury trials, minimized the powerful situational and systemic factors that engulfed these young men and women. Their actions were assumed to be products of free will, rational choice, and personal accountability. I argue not so when deindividuation, group mind, and the host of stress, exhaustion, sleep deprivation, and other psychological states are at play. They become transformed, just as the good angel, Lucifer, was transformed into the devil. Situations matter much more than most people realize or can acknowledge.

I've been teaching bright college students for nearly fifty years, and it's hard to get them to appreciate the situationist's analysis of evil, prejudice, or any kind of pathological behavior because our whole society is so wedded to the dispositional perspective: Good people do good deeds, and bad people do bad deeds. It's part of our institutional thinking. It's what psychiatry is all about. It's what medicine is all about. It's what the legal system is all about. And it's what religious systems are all about. We put good inside of people, and we put bad inside of people. It's so ingrained in the way we think, but the situationist's perspective says that although that may sometimes be true, we need to acknowledge that there can be powerful yet subtle social forces in given settings that have potentially transformative power over us.

That's why *Lord of the Flies* had such a big impact. How could it be that just changing your appearance could make you kill when such an activity was previously alien to you? That's still a very difficult message to get across. You can tell students that the majority of subjects in the Milgram experiment went all the way. How likely is it that you would do it? "Oh, no, I'm not that kind of person," they say. Well, the majority of guards in my study did brutal things. If you were a guard, what would you do? "I would be a good guard," they answer.

It's partly a self-serving bias. We want to believe we are good, we are different, we are better, or we are superior. But this body of social-psychological research—and there are obviously many more experiments in addition to mine and Milgram's—shows that the majority of good, ordinary, normal people can be easily seduced, tempted, or initiated into behaving in ways that they say they never would. In thirty minutes we got them stepping across that line. I don't know, but I would bet that if you went to more collectivist cultures, cultures that focus on the community or the group as the unit rather than the individual, they might buy into a situationist approach more freely.

The other important thing is to see this as a progression. If the Stanford prison study had continued on for three months then I'd have to predict that there would have been a steady increase in the level of dehumanization and degradation that might have rivaled the abuses at Abu Ghraib.

The other important thing in all of this is the "evil of inaction." I've been focusing on the perpetrators, but there are two important groups that I want to focus on more in my research and my future writing. What about all the people who observed what was happening and said nothing? There were doctors, nurses, and technicians. There is a photo in which two soldiers piled the prisoners up in a pyramid, and there were twelve other people standing around, watching. If you watch this happening and you don't say, "This is wrong! Stop it! This is awful!" you give tacit approval. You are the silent majority who makes something acceptable. If I get in a cab in New York and the cabdriver starts telling me a racist or sexist joke and I don't stop him, that means he will now tell that joke over and over again, thinking that his passengers like it. He takes my silence as approval of his racism. There is not only the evil of inaction among all those people in that prison, but also the people in society in general who observe evil and allow it to continue by not opposing it.

In our prison study it was the "good guards" who maintained the prison. It was the guards on the shift where you had the worst abuses who never did anything bad to the prisoners, but not once, over the whole week, did they ever go to one of the bad guards and say, "What are you doing? We get paid the same money without knocking ourselves out." Or, "Hey, remember those are college students, not prisoners." No good guard ever intervened once to stop the activities of the bad guards. No good guard ever came a minute late, left a minute early, or publicly complained. In a sense, then, it's the good guard who allows this to happen. It's the good parent who allows a spouse to abuse their children without opposing it. That's something that's really important for us to consider.

The other important group for us to recognize are the heroes in our midst. When you take a situationist approach you say the majority of people in these settings will go all the way and step across the line. Because evil is so fascinating, we have been obsessed with looking at evildoers. Well, what about the ones who didn't go all the way? We've ignored them, but those are, by definition, the heroes.

The hero is somebody who somehow has the inner qualities, inner resources, character, strength, or virtue—whatever you want to call it from Marty Seligman's Positive Psychology perspective—to resist those situational pressures. And we know nothing about those people. There has never been a psychology of heroism. For example, after the Holocaust it took thirty years before anyone asked the simple question of whether anybody helped the Jews. We were so obsessed with the evil of the Nazis that they didn't ask the question. When they asked, the answer was, Yes! In every country there were people who helped Jews. There were people who put their lives, and potentially the lives of their whole families, on the line to hide Jews in barns and attics when, if they were caught, they would be killed. Those are heroic deeds. When those people were interviewed years later, typically they said it was no big deal. They couldn't understand why other people didn't do it. It looked like they were a little more religious, but there is no research that studies the moment of decision when you are about to engage—to go along or to resist, to obey or to disobey. This is the kind of psychological research that would be exciting to do.

Philip Zimbardo

It can't ever be done again because all this research is now considered unethical, but in the case of Abu Ghraib we have a hero. A reserve specialist, a low-level guy, saw these pictures on a CD that his buddy gave him. He immediately recognized that this was immoral and wrong for Americans to ever do. At first he slipped the CD containing the images under the door of a superior officer. And then, interestingly, the next day he owned up to it. He said, "I was the one who put it there. I think this is wrong. You should take some action." I talked to some military people who say that it took enormous internal fortitude to do that, because as an army reservist in the military police in that setting, you are the lowest form of animal life in the military. It's only because he personally showed the pictures that they couldn't disown the fact that the abuse was happening, although they tried.

The paradox is that he's an incredible hero who is now in hiding. He's under protective custody. Soldiers in his own battalion say he disgraced them. Apparently there are death threats against him. But this whistle-blower's deed stopped the abuse. There's no question that it would have gone on. It's only because there is graphic visual evidence of how horrible these deeds are that the abuse stopped and led to more than a half dozen investigations. Again, here is somebody who fascinates me, because he is the rare person we would all like to imagine that we would be.

We like to think we're good, and down deep we'd all like to say, "I would be the heroic one. I would be the one who would blow the whistle." The limit of the situationist approach comes when we see these heroes, because it appears that somehow they have something in them that the majority doesn't. We don't know what that special quality is. Certainly it's something we want to study. We want to be able to identify it so we can nurture it and teach it to our children and to others in our society.

* * *

It's also very important for me, as somebody who's been interested in prisons for a very long time, to make sure that we don't see Abu Ghraib as an exotic Middle Eastern prison that is the only place where these terrible things happen. It may not be as extreme, but terrible things happen in prisons in our own country.

Right now there is an investigation of deaths of American Indians in prisons on Indian reservations. There is a large number of what the media calls "strange" circumstances of death. There's another investigation at the California Youth Authority concerning adolescents who were put in animal cages and drugged over long periods of time so that it would be easier for the guards to control them. At Pelican Bay, a maximum-security prison in California, prisoners are put in extreme solitary confinement units for twenty-three and a half hours a day for five or ten years on end in which they never get to see another human being.

One of the issues that society has to face is, if we ever let any of these people out, are they likely to be more or less dangerous to society than when they were put in? The answer is obvious. If you put kids in animal cages and drug them, when they get out they're going to be more like animals. If you put people in solitary confinement where they never interact with other people, how are they going to be able to interact on any normal basis when they get out? Prisons have lost any semblance of being places of rehabilitation. They are places of punishment and abuse.

Society never gave guards or prison administrators permission to do these things. Society says that if somebody breaks a law we want to separate him from the community for a certain period of time, period. That's what a sentence is. It's only in capital cases that we go one step further and kill him. We don't say that convicts should be abused, or should be put in places that degrade and dehumanize them, because we want them to come out and not go back. The fact that recidivism rates at most prisons are 60 percent and higher is evidence that this system doesn't work. In addition, the second offense is more likely to be more severe than the first. This means that prisons are places that breed crime and evil. This is not what society wants.

The bottom line is that nobody really cares what happens in prison. Nobody wants to know. Prisons are the default value of every society. We just want to dump convicts there, and let them come back and be good people. We only care about rapists and child molesters, so we want to keep track of them when they get out. For everybody else we don't want to know. We assume they go to prison, we'd like to believe they get rehabilitated, and when they come back they work in society. But from everything I know, most prisons are places that abuse prisoners, making them worse. They make them hate, make them want to get back at the injustice they've experienced.

All prisons are cloaked in a veil of secrecy. No one knows what happens in a prison. And when I say no one outside the prison knows, I mean mayors don't know, governors don't know, presidents don't know, and congressional subcommittees don't know. Prisons are huge places, and if you just walk in you wouldn't know what to see. They could direct you to one part of the prison where everything is clean and rosy and nice, and the prisoners are eating steak for your visit. Prisons have to lift the veil of secrecy. The media and lawyers have to have access to prisons.

None of that was true in Abu Ghraib. After a while the Red Cross was prohibited from coming in, because the Red Cross issued reports months before any of these exposés. Once you have a prison shrouded in secrecy, everyone in the prison knows there's no accountability outside. And then again, just as in my study of vandalism in the Bronx, nobody cares. Once you have that mentality as a guard or a prison administrator you say, "We don't really care what happens in prison, as long as it doesn't go too far. We don't want the prisoners to be killed. We don't want them to be extremely tortured. And obviously we don't want pictures of it thrust in our face on the evening news while we're having dinner. But short of that . . ."

And it's true. Now, one of the counterreactions to Abu Ghraib is that by forcing us to look at these things the whistle-blower almost becomes the culprit. We knew torture was going on at some level, and we knew it was important to get information from these bad people, but we just didn't want to know how bad it was, how far the soldiers had gone, and how far over the line they had stepped.

That's really something that every American citizen has to think about. Prisons are our property. We pay for everything in that prison. We pay all the guards' salaries, the superintendent's salary, and the warden's salary. The whole thing comes out of our tax dollars and we have to care. If the money is not going to rehabilitate prisoners, that means that when these people come out they're going to be attacking, murdering, stealing from us, doing all these terrible things again and again. Who wants to keep paying for that?

Not only do we have to care about what we do in prisons in America, but we also have to care what Americans do in prisons anywhere in the world that we run, because it sends not only a political message, it sends a moral message. One of the worst things about Abu Ghraib is that we have totally lost any sense of moral superiority that America ever had. Those pictures will be with the world for decades to come. We can say that we're bringing freedom and democracy to the world, but when people look at the pictures they say, "Yeah, and what else are you bringing?"

10

The Neurology of Self-Awareness

V. S. Ramachandran

Neuroscientist; Director, Center for Brain and Cognition, University of California, San Diego; Author, Phantoms in the Brain *and* The Tell-Tale Brain

What is the self? How does the activity of neurons give rise to the sense of being a conscious human being? Even this most ancient of philosophical problems, I believe, will yield to the methods of empirical science. It now seems increasingly likely that the self is not a holistic property of the entire brain; it arises from the activity of specific sets of interlinked brain circuits. But we need to know which circuits are critically involved and what their functions might be. It is the "turning inward" aspect of the self—its recursiveness—that gives it its peculiar paradoxical quality.

It has been suggested by Horace Barlow, Nick Humphrey, David Premack, and Marvin Minsky (among others) that consciousness may have evolved primarily in a social context. Minsky speaks of a second parallel mechanism that has evolved in humans to create representations *of* earlier representations, and Humphrey has argued that our ability to introspect may have evolved specifically to construct meaningful models of other people's minds in order to predict their behavior. "I feel jealous in order to understand what jealousy feels like in someone else"—a shortcut to predicting that person's behavior.

Here I develop these arguments further. If I succeed in seeing any further it is by "standing on the shoulders of these giants." Specifically, I suggest that "other awareness" may have evolved first and then counterintuitively, as often happens in evolution, the same ability was exploited to model one's own mind—what one calls self-awareness. I will also suggest that a specific system of neurons called mirror neurons are involved in this ability. Finally, I discuss some clinical examples to illustrate these ideas and make some testable predictions.

There are many aspects of self. It has a sense of unity despite the multitude of sense impressions and beliefs. In addition it has a sense of continuity in time, of being in control of its actions ("free will"), of being anchored in a body; a sense of its worth, dignity, and mortality (or immortality). Each of these aspects of self may be mediated by different centers in different parts of the brain, and it's only for convenience that we lump them together in a single word.

As noted earlier there is one aspect of self that seems stranger than all the others—the fact that it is aware of itself. I would like to suggest that groups of neurons called mirror neurons are critically involved in this ability.

The discovery of mirror neurons was made by Giacomo Rizzolatti, Vittorio Gallase, and Marco Iaccoboni while recording from the brains of monkeys that performed certain goal-directed voluntary actions. For instance, when the monkey reached for a peanut, a certain neuron in its premotor cortex (in the frontal lobes) would fire. Another neuron would fire when the monkey pushed a button, a third neuron when he pulled a lever. The existence of such command neurons that control voluntary movements has been known for decades. Amazingly, a subset of these neurons had an additional peculiar property. The neuron fired not only (say) when the monkey reached for a peanut, but also when it watched *another* monkey reach for a peanut! These were dubbed "mirror neurons" or "monkey-see-monkey-do" neurons. This was an extraordinary observation because it implies that the neuron (or more accurately, the network which it is part of) was not only generating a highly specific command ("reach for the nut") but was capable of adopting another monkey's point of view. It was doing a sort of internal virtual reality simulation of the

V. S. Ramachandran

other monkey's action in order to figure out what he was "up to." It was, in short, a "mind-reading" neuron.

Neurons in the anterior cingulate will respond to the patient being poked with a needle; they are often referred to as sensory pain neurons. Remarkably, researchers at the University of Toronto have found that some of them will fire equally strongly when the patient watches someone *else* being poked. I call these "empathy neurons" or "Dalai Lama neurons," for they are dissolving the barrier between self and others. Notice that in saying this one isn't being metaphorical; the neuron in question simply doesn't know the difference between it and others.

Primates (including humans) are highly social creatures, and knowing what someone is "up to"—creating an internal simulation of his/her mind—is crucial for survival, earning us the title "the Machiavellian primate." In an essay for *Edge* (2001) entitled "Mirror Neurons and the Great Leap Forward," I suggested that in addition to providing a neural substrate for figuring out another person's intentions (as noted by Rizzolatti's group), the emergence and subsequent sophistication of mirror neurons in hominids may have played a crucial role in many quintessentially human abilities, such as empathy, learning through *imitation* (rather than trial and error), and the rapid transmission of what we call "culture." (And "the great leap forward"—the rapid Lamarckian transmission of "accidental," one-of-a-kind inventions.)

I turn now to the main concern of this essay—the nature of self. When you think of your own self, what comes to mind? You have a sense of "introspecting" on your own thoughts and feelings and of "watching" yourself going about your business—*as if* you were looking at yourself from another person's vantage point. How does this happen?

Evolution often takes advantage of preexisting structures to evolve completely novel abilities. I suggest that once the ability to engage in cross-modal abstraction emerged—for example, between visual "vertical" on the retina and photoreceptive "vertical" signaled by muscles (for grasping trees) it set the stage for the emergence of mirror neurons in hominids. Mirror neurons are also abundant in the inferior parietal lobule—a structure that underwent an accelerated expansion in the great apes and, later, in humans. As the brain evolved further the lobule split into two gyri—the supramarginal gyrus that allowed you to "reflect" on your own anticipated actions, and the angular gyrus that allowed you to "reflect" on your body (on the right) and perhaps on other more social and linguistic aspects of your self (left hemisphere). I have argued elsewhere that mirror neurons are fundamentally performing a kind of abstraction across activity in visual maps and motor maps. This in turn may have paved the way for more conceptual types of abstraction, such as metaphor ("get a grip on yourself").

How does all this lead to self-awareness? I suggest that self-awareness is simply using mirror neurons for "looking at myself *as if* someone else is looking at me" (the word "me" encompassing some of my brain processes, as well). The mirror neuron mechanism—the same algorithm—that originally evolved to help you adopt *another's* point of view was turned inward to look at your own self. This, in essence, is the basis of things like "introspection." It may not be coincidental that we use phrases like "self-conscious" when you really mean that you are conscious of others being conscious of you. Or say "I am reflecting" when you mean you are aware of yourself thinking. In other words, the ability to turn inward to introspect or reflect may be a sort of metaphorical extension of the mirror neurons' ability to read other's minds. It is often tacitly assumed that the uniquely human ability to construct a "theory of other minds," or "TOM" (seeing the world from the other's point of view, "mind reading," figuring out what someone is up to, etc.), must come after an already pre-existing sense of self. I am arguing that the exact opposite is true:

V. S. Ramachandran

The TOM evolved *first* in response to social needs and then later, as an unexpected bonus, came the ability to introspect on your own thoughts and intentions. I claim no great originality for these ideas; they are part of the current zeitgeist. Any novelty derives from the manner in which I shall marshal the evidence from physiology and from our own work in neurology. Note that I am not arguing that mirror neurons are *sufficient* for the emergence of self; only that they must have played a pivotal role. (Otherwise, monkeys would have self-awareness, and they don't.) They may have to reach a certain critical level of sophistication that allowed them to build on earlier functions (TOM) and become linked to certain other brain circuits, especially the Wernicke's ("language comprehension") area and parts of the frontal lobes.

Does the mirror neuron theory of self make other predictions? Given our discovery that autistic children have deficient mirror neurons and correspondingly deficient TOM, we would predict that they would have a deficient sense of self (TMM) and difficulty with introspection. The same might be true for other neurological disorders; damage to the inferior parietal lobule/TPO junction (which are known to contain mirror neurons) and parts of the frontal lobes should also lead to a deficiency of certain aspects of self-awareness. (Incidentally, Gallup's mirror test—removing a paint splotch from your face while looking at a mirror—is not an adequate test of self-awareness, even though it is touted as such. We have seen patients who vehemently claim that their reflection in the mirror is "someone else," yet they pass the Gallup test!)

It has recently been shown that if a conscious, awake human patient has his parietal lobe stimulated during neurosurgery, he will sometimes have an "out of body" experience—as if he was a detached entity watching his own body from up near the ceiling. I suggest that this arises because of a dysfunction in the mirror neuron system in the parieto-occipital junction caused by the stimulating electrode.

These neurons are ordinarily activated when we temporarily "adopt" another's view of our body and mind (as outlined earlier in this essay). But we are always *aware* we are doing this partly because of other signals (both sensory and reafference/command signals) telling you you are not literally moving out of yourself. (There may also be frontal inhibitory mechanisms that stop you from involuntarily mimicking another person looking at you.)

If these mirror neuron–related mechanisms are deranged by the stimulating electrode the net result would be an out-of-body experience. Some years ago we examined a patient with a syndrome called anosognosia who had a lesion in his right parietal lobe and vehemently denied the paralysis. Remarkably the patient also denied the paralysis of another patient sitting in an adjacent wheelchair (who failed to move the arm on command from the physician)! Here again was evidence that two seemingly contradictory aspects of self—its individuation and intense privacy versus its social reciprocity—may complement each other and arise from the same neural mechanism, mirror neurons. Like the two sides of a Mobius strip, they are really the same, even though they appear—on local inspection—to be fundamentally different.

Have we solved the problem of self? Obviously not—we have barely scratched the surface. But hopefully we have paved the way for future models and empirical studies on the nature of self, a problem that philosophers have made essentially no headway in solving (and not for want of effort—they have been at it for three thousand years). Hence our grounds for optimism about the future of brain research—especially for solving what is arguably science's greatest riddle.

V. S. Ramachandran

11

Eudaemonia: The Good Life

Martin Seligman

Psychologist, University of Pennsylvania; Author, Authentic Happiness

In order to answer the question of what I want to do and what my ambitions are, it's worth surveying what psychology has done and what psychology can be proud of. The domain of psychology that I come from—clinical psychology, social psychology—has one major medal on its chest: If you look back to 1945 to 1950, no major mental illness was treatable. It was entirely smoke and mirrors. The National Institute of Mental Health essentially invested between $20 and $30 billion—It's never been the National Institute of Mental Health, by the way; it's always been Mental Illness—on the question of the relief of mental illness. And by my count, the $20 billion, fifty-year investment produced the following great achievements.

The first is that fourteen major mental illnesses are now treatable. Two of them are curable, either by specific forms of psychotherapy or specific drugs. The two curable ones—people always ask—are probably panic disorder and blood and injury phobia. So the first great thing that psychology and psychiatry did in our lifetime was to be able to relieve an enormous amount of suffering.

The second thing, which is even better from where I sit, is that a science of mental illness developed such that we found that we could measure fuzzy states like sadness, alcoholism, and schizophrenia with psychometric precision. We developed a classification, a *DSM*, so that people in London and in Philadelphia can agree that they're both seeing a bipolar depressive. Third, we are able to look at the causal skein of mental illness and unravel it, either by longitudinal studies—the same people over time—or experimental studies, which would get rid of third variables. Fourth, we're able to create treatments—drugs, psychotherapy—and do random-assignment placebo-control studies to find out which ones really worked and which ones were inert. That led to the following: that psychology and psychiatry can make miserable people less miserable. That's great. I'm all for it.

But there were three serious costs to selling out to the disease model—and it was a sellout, by the way; it was financial. In 1946 the Veterans Administration Act was passed, and practitioners found they could get jobs if they worked on mental illness, and that's what happened to the practice community. In 1947 NIMH was founded and academics like me found that you could get grants if you're working on mental illness. That's what happened to 90 percent of the science in psychology.

But there were three costs of becoming part of the disease model: The first one was moral, that we became victimologists and patholiogizers. Our view of human nature was that mental illness fell on you like a ton of bricks, and we forgot about notions like choice, responsibility, preference, will, character, and the like. The second cost was that by working only on mental illness we forgot about making the lives of relatively untroubled people happier, more productive, and more fulfilling. And we completely forgot about genius, which became a dirty word. The third cost was that because we were trying to undo pathology we didn't develop interventions to make people happier; we developed interventions to make people less miserable.

That's the background. What's missing is the question of whether psychologists can make people lastingly happier. That is, can we apply the same kind of scientific method to get cumulative, replicable interventions? I'm interested in psychological ones, but an obvious question applies to pharmacology—not to take people from –8 to –5, but to take people from +2 to +6. My great ambition for psychology, and I hope to play a role in it, is that in the next ten to fifteen years we will be able to make the parallel claim about happiness; that is, in the same way I can claim unblushingly that psychology and psychiatry have decreased the tonnage of suffering in the world, my aim is that psychology and maybe psychiatry will increase the tonnage of happiness in the world.

Happiness is a hopelessly vague shorthand for other things, so when I began to work in positive psychology, my first task was a wheat/chaff task to try to say what the measurable components of what people mean by happiness are. What are the workable pieces of it? The word "happiness," like the word "cognition," plays no role in cognitive theory. Cognition is about memory, perception, etc. The field of happiness is about other things. The word "happiness" plays only a role for labeling what we're doing.

What's workable within happiness are three different kinds of lives: The first is the pleasant life, which consists of having as many of the positive emotions as you can, and learning the skills that amplify them. There are a half dozen such skills that have been reasonably well documented. That's the Hollywood view of happiness, the Debbie Reynolds, smiley, giggly view of happiness. It's positive emotion. But, one might ask, isn't that where positive psychology ends? Isn't pleasure all there is to the positive side of life? You only have to look superficially back to the history of philosophy to find out that from Aristotle through Seneca through Wittgenstein the notion of pleasure was thought of as vulgar. There's very good intellectual provenance for two other kinds of happy lives, which in the Hollywood/American conception have gone by the boards. Part of my job is to resurrect them.

The second one is eudaemonia, the good life, which is what Thomas Jefferson and Aristotle meant by the pursuit of happiness. They did not mean smiling a lot and giggling. Aristotle talks about the pleasures of contemplation and the pleasures of good conversation. Aristotle is not talking about raw feeling, about thrills, about orgasms. Aristotle is talking about what Mike Csikszentmihalyi works on, and that is, when one has a good conversation, when one contemplates well. When one is in eudaemonia, time stops. You feel completely at home. Self-consciousness is blocked. You're one with the music.

The good life consists of the roots that lead to flow. It consists of first knowing what your signature strengths are and then recrafting your life to use them more—recrafting your work, your romance, your friendships, your leisure, and your parenting to deploy the things you're best at. What you get out of that is not the propensity to giggle a lot; what you get is flow, and the more you deploy your highest strengths the more flow you get in life.

Coming out this month as part of the *DSM* is a classification of strengths and virtues; it's the opposite of the classification of the insanities. When we look we see that there are six virtues, which we find endorsed across cultures, and these break down into twenty-four strengths. The six virtues that we find are nonarbitrary—first, a wisdom and knowledge cluster; second, a courage cluster; third, virtues like love and humanity; fourth, a justice cluster; fifth, a temperance, moderation cluster; and sixth, a spirituality, transcendence cluster. We sent people up to northern Greenland, and down to the Masai, and are involved in a seventy-nation study in which we look at the ubiquity of these. Indeed, we're beginning to have the view that those six virtues are just as much a part of human nature as walking on two feet are.

I can give you some examples of what I mean by recrafting your life to use your signature strength and getting flow. One person I worked with was a bagger at Genuardi's. She didn't like bagging, took the signature strengths test, and her highest strength was social intelligence. And so she recrafted her job to make the encounter with her the social highlight of every customer's day. She obviously failed at that a lot, but by deploying the single thing she was best at, she changed the job from one in which time hung heavy on her hands into one in which time flew by.

So just to review so far, there is the pleasant life—having as many of the pleasures as you can and the skills to amplify them—and the good life—knowing what your signature strengths are and recrafting everything you do to use them as much as possible. But there's a third form of life, and if you're a bridge player like me, or a stamp collector, you can have eudaemonia; that is, you can be in flow. But everyone finds that as they grow older and look in the mirror they worry that they're fidgeting until they die. That's because there's a third form of happiness that is ineluctably pursued by humans, and that's the pursuit of meaning. I'm not going to be sophomoric enough to try to tell *Edge* readers the theory of meaning, but there is one thing we know about meaning: that meaning consists in attachment to something bigger than you are. The self is not a very good site for meaning, and the larger the thing that you can credibly attach yourself to, the more meaning you get out of life.

There's an enormous range of things that are larger than us that we can belong to and be part of, some of which are prepackaged. Being an Orthodox Jew, for example, or being a Republican are prepackaged ones. Being a teacher, someone whose life is wrapped up in the growth of younger people, is a non-prepackaged one. Being an agent is a non-prepackaged one—it's a life in service of the people you conceive of to be the greatest minds on the planet. And they wouldn't do their thing without agency. You can convert agency into the idea that "I'm just doing it for all the money I make," and then it's not a meaningful life. But I don't think you wake up in the morning raring to make more money; it's rather in service of this much larger goal of the intellectual salon. Being a lawyer can either be a business just in service of making a half million dollars a year, in which case it's not meaningful, or it can be in service of good counsel, fairness, and justice. That's the non-prepackaged form of meaning.

Aristotle said the two noblest professions are teaching and politics, and I believe that as well. Raising children, and projecting a positive human future through your children, is a meaningful form of life. Saving the whales is a meaningful form of life. Fighting in Iraq is a meaningful form of life. Being an Arab terrorist is a meaningful form of life.

Notice, this isn't a distinction between good and evil. That's not part of this. This isn't a theory of everything. This is a theory of meaning, and the theory says joining and serving in things larger than you that you believe in while using your highest strengths is a recipe for meaning. One of the things people don't like about my theory is that suicide bombers and the firemen who saved lives and lost their lives both had meaningful lives. I would condemn one as evil and the other as good, but not on the grounds of meaning.

Within the psychological community, there are two different ways that this plays out. First, let me contrast the therapeutic model to the coaching model. The therapeutic model involves fixing broken things. Ten years ago when I introduced myself to my seatmate on an airplane, and they asked me what I did, and I told them I was a psychologist, they'd move away from me. That's because they actually had the right idea: that the job of a psychologist is to find out what's really wrong with you. Now when I tell people that I work on positive psychology they move toward me. That's because the job of the positive psychologist is to find out what's really right with you—something you may not be aware of—and to get you to use it more and more.

I'm reluctant to use the words "paradigm shift," and I'm reluctant to use the word "school," or "movement," but here's what I can say empirically: I chart the growth of this approach in a number of ways. It's gone from an endeavor in which seven years ago there were no courses in the United States on positive psychology; now there are a couple of hundred, at many major universities. I teach positive psy-

Martin Seligman

chology at the introductory level. I've raised $30 million in the last few years for the scientific infrastructure of positive psychology. Like many scientists I've spent my life on my knees as a supplicant to one agency or another, but it's never been so easy for me to raise money in my life. I've never found a situation before in which people will come up after a speech and write me a check and say, "Do something good with it." This is something that rings a bell as far as support goes.

I decided a year ago that the time had come to disseminate some of this. We had a good six or seven years of scientific discovery behind us, so I started to disseminate it to the disseminators. I now teach a course to 550 professionals every Wednesday on the telephone. It's the largest conference call ever given. It's a six-month course which consists of clinical psychologists, coaches, CEOs, and personnel managers from twenty countries and virtually every state. We gather once a week on the phone, and I give a one-hour master class. At the end of each master class I give an intervention like a gratitude visit, or taking the signature strengths test, or writing your vision of a positive human future, and then you do the exercise yourself and with your clients. You measure levels of happiness before and after, and then once a week you meet on the telephone in groups of about fifteen with a master clinician to go over it.

We've gone from 0 to 550 people in a year who have brought this stuff into their practices. It's my ambition that we will find out what works and what doesn't work. That is, we will go through the same random-assignment placebo-control procedures. I've gathered over a hundred interventions that have been claimed, from the Buddha to Tony Robbins, to make people happier. My guess is that 90 percent of these are inert.

If you go to my Web site, www.authentichappiness.org, and you take a variety of happiness and depression tests, you can then go to a link called Interventions. This link says that we want to find out what really works, and to do that we're going to randomly assign you to an intervention. You won't know if it's a placebo or not. And then you will carry out this intervention, and you will journal it, and then we will follow you for the next year. We've now done this with about six different interventions. I'm not going to give away the placebo, but here is one non-placebo:

About 300 people have gone through the gratitude visit. In the gratitude visit—and as you're reading this I'd like everyone to do it—you think of someone in your life who made an enormous positive difference, who's still alive, whom you never properly thanked. You've got such a person? It's important to be able to do that, by the way, since empirically the amount of gratitude is related to baseline levels of happiness. The less gratitude you have in life the more unhappy you are, interestingly.

If you were going to make a gratitude visit, you do the following: First you'd write a 300-word testimonial to that person; concrete, well-written, telling the story of what they did, how it made a difference, and where you are in life now as a result. Then you'd call him up and say, "I want to come visit you." And he'd ask you why, and you'd say, "I don't want to tell you. It's a surprise." And you'd show up at his door, sit down, and read the testimonial—it turns out everyone weeps when this happens—and then a week later, a month later, three months later, and a year later, we give you the battery of tests, and ask the question relative to placebo controls, "Are you happier? Are you less depressed?" It turns out the gratitude visit is one of the exercises which, to my surprise, makes people lastingly less depressed and happier than the placebo.

If you consider EST, or Tony Robbins, or the Maharishi, these are not dumb people. They've invented a lot of interventions. Tony Robbins has people doing fire walks; EST, I gather, has people not going to the bathroom for twenty-four hours, and the like. Some of these actually work, and some don't. The challenge is to subject them to the nasty thumb of science. A great deal of my work now is to take all of these interventions, manualize them, randomly assign people to them, and then look to see if in the long run these make people lastingly happier. My ambition and my optimism for psychology over the next fifteen years is that we will actually have a set of interventions which will reliably make people happier, and many of which you can do yourself. You won't need to go to therapists to do it. The method for finding out what works is the same old method; that is, the random-assignment placebo-controlled study that we did with misery. It's exactly the same question for making people happier.

I'm not going to give away a placebo, but let me just say a couple of things about it. It turns out we've already found out that several of the things that have been proposed—from the Buddha to Tony Robbins—don't work. We've got them up there on the Web site, people do them, and we find that there's no lasting change in either lowering depression or raising the level of happiness. But they're plausible; they're things that you or I would think would work, but because some of your readers are now going to jump to www.authentic happiness.org and get into the placebo, I don't want to give away what the placebos are. The interesting thing is that some of these things actually lastingly make people happier, and others don't. The aim of science is to find out what the active ingredients are.

I spent the first thirty years of my career working on misery. The first thing I worked on was learned helplessness. I found helpless dogs, helpless rats, and helpless people, and I began to ask, almost forty years ago now, how do you break it up? What's the neuroscience of it? What drugs work? While working on helplessness there was a finding I was always brushing under the rug, which was that with people and with animals, when we gave them uncontrollable events, only five out of eight became helpless. About a third of them we couldn't make helpless. And about a tenth of them were helpless to begin with and we didn't have to do anything.

About twenty-five years ago I began to ask the question, who never gets helpless? That is, who resists collapsing? And the reverse question is, who becomes helpless at the drop of a hat? I got interested in optimism because I found out that the people who didn't become helpless were people who when they encountered events in which nothing they did mattered, thought about those events as being temporary, controllable, local, and not their fault; whereas people who collapsed in a heap immediately upon becoming helpless were people who saw the bad event as being permanent, uncontrollable, pervasive, and their fault. Twenty-five years ago I started working on optimism versus pessimism, and I found that optimistic people got depressed at half the rate of pessimistic people; that optimistic people succeeded better in all professions that we measured except one; that optimistic people had better, feistier immune systems, and probably lived longer than pessimistic people. We also created interventions that reliably changed pessimists into optimists.

That's what I did up until about six years ago. Six or seven years ago I decided I was going to run for president of the American Psychological Association, and I got elected by the largest margin in the history of the association—to my surprise, since I'm not at all a political person. They told me after I was elected that presidents are supposed to have themes, initiatives. I didn't know what mine was going to be. I thought my initiative might be prevention, since I knew a lot about prevention, so I gathered together the twelve leading people in the world from prevention. We met for a day asking the question, could the prevention of mental illness be a presidential initiative? I have to confess to you that I have the attention span of an eight-year-old, but this was really boring. They basically said, "Let's take the things that work on schizophrenia and do them earlier in life." As I was walking out with Mihaly Csikszentmihalyi, he said, "Marty, this has no intellectual backbone. You've got to do something better than this."

Two weeks later I had an epiphany. It changed my life, and I hope it's changed the course of psychology. I was in my garden with my five-year-old daughter, Nicky, and to make another confession, even though I've written a book about children and have worked with children, I'm no good with them since I'm time-urgent and task-oriented. I was weeding, and Nicky was throwing weeds into the air, dancing, singing, and having a wonderful time—and I shouted at her. She walked away, puzzled, and walked back and said, "Daddy, I want to talk to you."

I said, "Yeah, Nicky?"

And she said, "Daddy, do you remember before my fifth birthday"—she had turned five about two weeks before—"I was a whiner? That I whined every day?"

I said, "Yeah, I remember—you were a horror."

"Have you noticed since my fifth birthday, Daddy, I haven't whined once?"

"Yeah, Nicky."

And she said, "Daddy, on my fifth birthday I decided I wasn't going to whine anymore. And that was the hardest thing I've ever done. And if I can stop whining, you can stop being such a grouch." In that moment, three things happened to me. The first was I realized that Nicky was right about me, that I had spent more than fifty years being a nimbus cloud and I didn't have a theory about why it's good to be a grouch. Some people talk about depressive realism, the idea that depressed people see reality better, but it occurred to me that maybe any success I'd had in life was in spite of being a grouch, not because of being a grouch, so I resolved to change. You haven't known me long enough, but people who have known me for that span of time know that I'm a sunnier person and deploy my critical intelligence less. I'm better able to see what's right, and I'm better at suppressing my falcon-like vigilance for what's wrong.

The second part of the epiphany was that I realized that my theories of child rearing were wrong. The theories of child rearing that the last two generations have been raised with in psychology are remedial. They basically say the job of the parent is to correct the kid's errors, and somehow out of the correction of errors an exemplary child rises. But if you think about Nicky, she corrected her own error, and my job was to take this extraordinary strength she had just shown, see into the soul, name it—social intelligence—help her to live her life around it, and to use it as a buffer against troubles. If you think about your own life, your success has not been because you've corrected your weaknesses, but because you found out a couple of things you were really good at, and you used those to buffer you against troubles. So the second thing I realized was that with any program whose aim is to correct what's wrong, even if it's asymptotically successful, the best it can ever get to is zero. And yet when you lie in bed at night you're not thinking about how to go from –5 to –2; you're generally thinking about how to go from +2 to +6 in life. It was interesting to me that there was no science for that. All of the science was remedial, correcting the negatives.

That led to the third, final, and most important part of the epiphany: I realized that my profession in social science generally was half-baked. The part that was baked was about victims, suffering and trauma, depression, anxiety, anger, and on and on. I'd spent my life on that and we knew a lot about it. That's what I meant by saying that the medal on our chest is that we can make miserable people less miserable. But the part that was unbaked was about what makes life worth living. What is happiness? What is virtue? What is meaning? What is strength? How are these things built? It became my mission in life, from that moment in the garden, to help to create a positive psychology whose mission would be the understanding and building of positive emotion, of strength and virtue, and of positive institutions.

I've spent a fair amount of my life asking questions about drugs and psychotherapy and their effects. Let me tell you how I summarize their effectiveness and then what I think the implications of that are for positive psychology.

First, it's important to know that in general there are two kinds of medications. There are palliatives, cosmetics like quinine for malaria, which suppress the symptoms for as long as you take them; when you stop taking quinine, the malaria returns at full force. Then there are curative drugs, like antibiotics for bacterial infection. When you stop taking those the bacteria are dead and don't recur.

The dirty little secret of biological psychiatry is that every single drug in the psychopharmacopia is palliative. That is, all of them are symptom suppressors, and when you stop taking them you're back at square one. In general for depression, for example, seratonin and the earlier tricyclic antidepressants work about 65 percent of the time. Interestingly, the two major forms of psychotherapy for depression—cognitive therapy and interpersonal therapy—are a tie. They work about 65 percent of the time. The difference, interestingly, is on relapse and recurrence. In interpersonal and cognitive therapy you actually learn a set of skills that you remember, so three years later when depression comes back you can start disputing catastrophic thoughts again. But if you had seratonin, or tricyclic antidepressants, three years later when it comes back it comes back in full force.

So that's part one—that the psychoactive drugs are palliative only, not curative. And lord knows I'm not a Freudian, but the thing I like best about Freud is that he was interested in cure. He was interested in antibiotics. He wasn't interested in palliation; indeed, that's what the whole displacement symptoms substitution is about. Biological psychiatry and psychology need to rediscover the question of cure. That's one of the reasons that I'm interested in positive psychology. When I told the Nicky story I talked about buffering against the troubles with the strengths; that's the kind of thing that lifts us to about the 65 percent barrier. That is, skilled clinicians often tell me that they've worked to bring out people's strengths, but never learned how to do it in graduate school. Part of what I'm training people to do is how to systematically test for the strengths, build them, and use them as buffers.

What are the "therapeutic" and drug prospects for positive psychology? Pleasant life, pleasures; good life, flow; meaningful life. And each of these I think has different possibilities. There are psychological interventions that I believe are effective for all three of those—indeed, that's what I meant by the random-assignment placebo-control endeavor. The question is, are we likely to find drugs that work on the pleasant life, the good life, and the meaningful life?

The answer is probably yes for the pleasant life. That is, there's a neuroscience that's relevant to the positive emotions, and people like Richard Davidson are beginning to pin down some localization within the brain. There are also recreational drugs—antidepressants don't bring pleasure, but recreational drugs do. I've never taken Ecstasy or cocaine, but I gather that they work on pleasure as well. At any rate, a pharmacology of pleasure is not science fiction, and I expect that as positive psychology matures our drug company friends will get interested in it. There are shortcuts to pleasure, and if you play with the relevant neural circuits, those are shortcuts.

Flow, however, doesn't have shortcuts. When I was an undergraduate one of my teachers, Julian Jaynes, a peculiar but wonderful man, was a research associate at Princeton. Some people said he was a genius; I didn't know him well enough to know. He was given a South American lizard as a laboratory pet, and the problem about the lizard was that no one could figure out what it ate, so the lizard was dying. Julian killed flies, and the lizard wouldn't eat them; blended mangoes and papayas, the lizard wouldn't eat them; Chinese takeout, the lizard had no interest. One day Julian came in and the lizard was in torpor, lying in the corner. He offered the lizard his lunch, but the lizard had no interest in ham on rye. He read the *New York Times* and he put the first section down on top of the ham on rye. The lizard took one look at this configuration, got up on its hind legs, stalked across the room, leapt up on the table, shredded the *New York Times*, and ate the ham sandwich. The moral is that lizards don't copulate and don't eat unless they go through the lizardly strengths and virtues first. They have to hunt, kill, shred, and stalk. And while we're a lot more complex than lizards, we have to as well. There are no shortcuts for us to reach flow. We have to indulge in our highest strengths in order to get eudaemonia. So can there be a shortcut? Can there be a pharmacology of it? I doubt it.

The third form of happiness, which is meaning, is again knowing what your highest strengths are and deploying those in the service of something you believe is larger than you are. There's no shortcut to that. That's what life is about. There will likely be a pharmacology of pleasure, and there may be a pharmacology of positive emotion generally, but it's unlikely there'll be an interesting pharmacology of flow. And it's impossible that there'll be a pharmacology of meaning.

12

What Are Numbers, Really? A Cerebral Basis For Number Sense

Stanislas Dehaene

Neuroscientist, Collège de France, Paris; Author, The Number Sense *and* Reading in the Brain

In a recent book as well as in a heated discussion on *Edge*, the mathematician Reuben Hersh has asked "What is mathematics, really?" This is an age-old issue that was already discussed in Ancient Greece and that puzzled Einstein twenty-three centuries later. I personally doubt that philosophical inquiry alone will ever provide a satisfactory answer (we don't even seem to be able to agree on what the question actually means!). However, if we want to use a scientific approach, we can address more focused questions, such as where specific mathematical objects like sets, numbers, or functions come from, who invented them, to what purpose they were originally put to use, their historical evolution, how are they acquired by children, and so on. In this way, we can start to define the nature of mathematics in a much more concrete way that is open to scientific investigation using historical research, psychology, or even neuroscience.

This is precisely what a small group of cognitive neuropsychologists in various countries and myself have been seeking to do in a very simple area of mathematics, perhaps the most basic of all: the domain of the natural integers 1, 2, 3, 4, etc. Our results, which are now based on literally hundreds of experiments, are quite surprising: Our brain seems to be equipped from birth with a number sense. Elementary arithmetic appears to be a basic, biologically determined ability inherent in our species (and not just our own—since we share it with many animals). Furthermore, it has a specific cerebral substrate, a set of neuronal networks that are similarly localized in all of us and that hold knowledge of numbers and their relations. In brief, perceiving numbers in our surroundings is as basic to us as echolocation is to bats or birdsong is to songbirds.

It is clear that this theory has important, immediate consequences for the nature of mathematics. Obviously, the amazing level of mathematical development that we have now reached is a uniquely human achievement, specific to our language-gifted species, and largely dependent on cultural accumulation. But the claim is that basic concepts that are at the foundation of mathematics, such as numbers, sets, space, distance, and so on, arise from the very architecture of our brain.

In this sense, numbers are like colors. You know that there are no colors in the physical world. Light comes in various wavelengths, but wavelength is not what we call color (a banana still looks yellow under different lighting conditions, where the wavelengths it reflects are completely changed). Color is an attribute created by the V4 area of our brain. This area computes the relative amount of light at various wavelengths across our retina, and uses it to compute the reflectance of objects (how they reflect the incoming light) in various spectral bands. This is what we call color, but it is purely a subjective quality constructed by the brain. It is, nonetheless, very useful for recognizing objects in the external world, because their color tends to remain constant across different lighting conditions, and that's presumably why the color perception ability of the brain has evolved in the way it has.

My claim is that number is very much like color. Because we live in a world full of discrete and movable objects, it is very useful for us to be able to extract number. This can help us to track predators or to select the best foraging grounds, to mention only very obvious examples. This is why evolution has endowed our brains and those of many animal species with simple numerical mechanisms. In animals, these mechanisms are very limited, as we shall see below: They are approximate, their representation becomes coarser for increasingly large numbers, and they involve only the simplest arithmetic operations (addition and subtraction). We humans have also had the remarkable good fortune to develop abilities for language and for symbolic notation. This has enabled us to develop exact mental representations for large numbers, as well as algorithms for precise calculations. I believe that mathematics, or at least arithmetic and number theory, is a pyramid of increasingly more abstract mental constructions based solely on (1) our ability for symbolic notation, and (2) our nonverbal ability to represent and understand numerical quantities.

So much for the philosophy now, but what is the actual evidence for these claims? Psychologists are beginning to realize that much of our mental life rests on the operation of dedicated, biologically determined mental modules that are specifically attuned to restricted domains of knowledge, and that have been laid down in our brains by evolution (see Steve Pinker's *How the Mind Works*). For instance, we seem to have domain-specific knowledge of animals, food, people, faces, emotions, and many other things. In each case—and number is no exception—psychologists demonstrate the existence of a domain-specific system of knowledge using the following four arguments:

- One should prove that possessing prior knowledge of the domain confers an evolutionary advantage. In the case of elementary arithmetic, this is quite obvious.
- There should be precursors of the ability in other animal species. Thus, some animals should be shown to have rudimentary arithmetic abilities. There should be systematic parallels between their abilities and those that are found in humans.

- The ability should emerge spontaneously in young children or even infants, independently of other abilities such as language. It should not be acquired by slow, domain-general mechanisms of learning.
- The ability should be shown to have a distinct neural substrate.
- My book *The Number Sense* is dedicated to proving these four points, as well as to exploring their consequences for education and for the philosophy of mathematics. In fact, solid experimental evidence supports the above claims, making the number domain one of the areas in which the demonstration of a biologically determined, domain-specific system of knowledge is the strongest. Here, I can only provide a few examples of experiments.

1. Animals have elementary numerical abilities. Rats, pigeons, parrots, dolphins, and of course primates can discriminate visual patterns or auditory sequences based on number alone (every other physical parameter being carefully controlled). For instance, rats can learn to press one lever for two events and another for four events, regardless of their nature, duration, and spacing, and whether they are auditory or visual. Animals also have elementary addition and subtraction abilities. These basic abilities are found in the wild, and not just in laboratory-trained animals. Years of training, however, are needed if one wants to inculcate number symbols into chimpanzees. Thus, approximate manipulations of numerosity are within the normal repertoire of many species, but exact symbolic manipulation of numbers isn't—it is a specifically human ability, or at least one that reaches its full-blown development in humans alone.

Stanislas Dehaene

2. There are systematic parallels between humans and animals. Animals' numerical behavior becomes increasingly imprecise for increasingly large numerals (number size effect). The same is true for humans, even when manipulating Arabic numerals: we are systematically slower to compute, say, 4+5 than 2+3. Animals also have difficulties discriminating two close quantities, such as 7 and 8. We too: when comparing Arabic digits, it takes us longer to decide that 9 is larger than 8 than to make the same decision for 9 versus 2 (and we make more errors, too).

3. Preverbal human infants have elementary numerical abilities, too. These are very similar to those of animals: Infants can discriminate two patterns based solely on their number, and they can make simple additions and subtractions. For instance, at five months of age, when one object is hidden behind a screen, and then another is added, infants expect to see two objects when the screen drops. We know this because careful measurements of their looking times show that they look longer when a trick makes a different number of objects appear. Greater looking time indicates that they are surprised when they see impossible events such as 1+1=1, 1+1=3, or 2−1=2. (Please, even if you are skeptical, don't dismiss these data with the back of your hand, as I was dismayed to discover Martin Gardner was doing in a recent review of my book for the *Los Angeles Times*. Sure enough, "measuring and averaging such times is not easy," but it is now done under very tightly controlled conditions, with double-blind videotape scoring. I urge you to read the original reports; for instance, Wynn, 1992, *Nature*, vol. 348, pp. 749–50—you'll be amazed at the level of

detail and experimental control that is brought to such experiments.)

Like animals and adults, infants are especially precise with small numbers, but they can also compute more approximately with larger numbers. In passing, note that these experiments, which are very reproducible, invalidate Piaget's notion that infants start out in life without any knowledge of numerical invariance. In my book, I show why Piaget's famous conservation experiments are biased and fail to tell us about the genuine arithmetical competence of young children.

4. Brain lesions can impair number sense. My colleagues and I have seen many patients at the hospital who have suffered cerebral lesions and, as a consequence, have become unable to process numbers. Some of these deficits are peripheral and concern the ability to identify words or digits or to produce them aloud. Others, however, indicate a genuine loss of number sense. Lesions to the left inferior parietal lobe can result in a patient remaining able to read and write Arabic numerals to dictation while failing to understand them. One of our patients couldn't do 3 minus 1, or decide which number fell between 2 and 4! He didn't have any problem telling us what month fell between February and April, however, or what day was just before Wednesday. Hence, the deficit was completely confined to numbers. The lesion site that yields such a number-sense deficit is highly reproducible in all cultures throughout the world.

5. Brain imaging during number processing tasks reveals a highly specific activation of the inferior parietal lobe, the very same region that, when lesioned, causes nu-

merical deficits. We have now seen this activation using most of the imaging methods currently available. PET scanning and fMRI pinpoint it anatomically to the left and right intraparietal sulci. Electrical recordings also tell us that this region is active during operations such as multiplication or comparison, and that it activates about 200 milliseconds following the presentation of a digit on a screen. There are even recordings of single neurons in the human parietal lobe (in the very special case of patients with intractable epilepsy) that show specific increases in activity during calculation.

The fact that we have such a biologically determined representation of number in our brain has many important consequences that I have tried to address in the book. The most crucial one is, of course, the issue of how mathematical education modifies this representation, and why some children develop a talent for arithmetic and mathematics while others (many of us!) remain innumerate. Assuming that we all start out in life with an approximate representation of number, one that is precise only for small numbers and that is not sufficient to distinguish 7 from 8, how do we ever move beyond that "animal" stage? I think that the acquisition of a language for numbers is crucial, and it is at that stage that cultural and educational differences appear. For instance, Chinese children have an edge in learning to count, simply because their number syntax is so much simpler. Whereas we say "seventeen, eighteen, nineteen, twenty, twenty-one, etc.," they say much more simply: "ten-seven, ten-eight, ten-nine, two-tens, two-tens-one, etc."; hence, they have to learn fewer words and a simpler syntax. Evidence indicates that the greater simplicity of their number words speeds up learning the counting sequence by about one year! But, I hasten to say, so does better organization in Asian classrooms, as shown by the UCLA psychologist Jim Stigler. As children move on to higher mathematics, there is considerable evidence that moving beyond approximation to learn exact calculation is very difficult for children and quite taxing even for the adult brain, and that strategies and education have a crucial impact.

Why, for instance, do we experience so much difficulty in remembering our multiplication tables? Probably because our brain never evolved to learn multiplication facts in the first place, so we have to tinker with brain circuits that are ill-adapted for this purpose (our associative memory causes us to confuse 8x3 with 8x4 as well as will 8+3). Sadly enough, innumeracy may be our normal human condition, and it takes us considerable effort to become numerate. Indeed, a lot can be explained about the failure of some children at school, and about the extraordinary success of some idiot savants in calculation, by appealing to differences in the amount of investment and in the affective state that they are in when they learn mathematics. Having reviewed much of the evidence for innate differences in mathematical abilities, including gender differences, I don't believe that much of our individual differences in math are the result of innate differences in "talent." Education is the key, and positive affect is the engine behind success in math.

The existence of mathematical prodigies might seem to go against this view. Their performance seems so otherworldly that they seem to have a different brain from our own. Not so, I claim—or at the very least, not so at the beginning of their lives: they start in life with the same endowment as the rest of us, a basic number sense, an intuition about numerical relations. Whatever is different in their adult brains is the result of successful education, strategies, and memorization. Indeed, all of their feats, from root extraction to multidigit multiplication, can be explained by simple tricks that any human brain can learn, if one were willing to make the effort.

Here is one example: the famous anecdote about Ramanujan and Hardy's taxi number. The prodigious Indian mathematician Ramanujan was slowly dying of tuberculosis when his colleague Hardy came to visit him and, not knowing what to say, made the following reflection: "The taxi that I hired to come here bore the number 1729. It seemed a rather dull number." "Oh, no, Hardy," Ramanujan replied, "it is a captivating one. It is the smallest number that can be expressed in two

At first sight, pital bed seems i sible. But in fact a the Indian mathe worked for decad rized scores of fac

1x1x1 = 1
2x2x2 = 8
3x3x3 = 27
4x4x4 = 64
5x5x5 = 125
6x6x6 = 216
7x7x7 = 343
8x8x8 = 512
9x9x9 = 729
10x10x10 = 1,000
11x11x11 = 1,331
12x12x12 = 1,728

Now, if you look at this list you see that (a) 1,728 is a cube; (b) 1,728 is one unit off 1,729, and 1 is also a cube; (c) 729 is also a cube; and (d) 1,000 is also a cube. Hence, it is absolutely *obvious* to someone with Ramanujan's training that 1,729 is the sum of two cubes in two different ways—namely, 1,728+1 and 1,000+729. Finding out that it is the smallest such number is more tricky, but can be done by trial and error. Eventually, the magic of this anecdote totally dissolves when one learns that Ramanujan had written this computation in his notebooks as an adolescent, and hence did not compute this on the spur of the moment in his hospital bed: he already knew it!

Would it be far-fetched to suggest that we could all match Ramanujan's feat with sufficient training? Perhaps that suggestion would seem less absurd if you consider that any high school student, even one who is not considered particularly bright, knows at least as much about mathematics as the most advanced mathematical scholars of the Middle Ages. We all start out in life with very similar brains, all endowed with an elementary number sense that has some innate structure, but also a degree of plasticity that allows it to be shaped by culture.

Back to the philosophy of mathematics, then. What are numbers, really? If we grant that we are all born with a rudimentary number sense that is engraved in the very architecture of our brains by evolution, then clearly numbers should be viewed as a construction of our brains. However, contrary to many social constructs such as art and religion, number and arithmetic are not arbitrary mental constructions. Rather, they are tightly adapted to the external world. Whence this adaptation? The puzzle about the adequacy of our mathematical constructions for the external world loses some of its mystery when one considers two facts:

- First, the basic elements on which our mathematical constructions are based, such as numbers, sets, space, and so on, have been rooted in the architecture of our brains by a long evolutionary process. Evolution has incorporated in our minds/brains structures that are essential to survival and hence to veridical perception of the external world. At the scale we live in, number is essential because we live in a world made of movable, denumerable objects. Things might have been very different if we lived in a purely fluid world, or at an atomic scale—and hence I concur with a few other mathematicians, such as Henri Poincare, Max Delbruck, or Reuben Hersh, in thinking that other life forms could have had mathematics very different from our own.

- Second, our mathematics has seen another evolution, a much faster one: a cultural evolution. Mathematical objects have been generated at will in the minds of mathematicians of the past thirty centuries (this is what we call "pure mathematics"). But then they have been selected for their usefulness in solving real-world problems, for instance in physics. Hence, many of our current mathematical tools are well adapted to the outside world, precisely because they were selected as a function of this fit.

Many mathematicians are Platonists. They think that the universe is made of mathematical stuff, and that the job of mathematicians is merely to discover it. I strongly deny this point of view. This does not mean, however, that I am a "social constructivist," as Martin Gardner would like to call me. I agree with Gardner, and against many social constructivists, that mathematical constructions transcend specific human cultures. In my view, however, this is because all human cultures have the same brain architecture that "resonates" to the same mathematical tunes. The value of pi, thank God, does not change with culture! (See the Sokal affair.) Furthermore, I am in no way denying that the external world provides a lot of structure, which gets incorporated into our mathematics. I only object to calling the structure of the universe "mathematical." We develop mathematical models of the world, but these are only models, and they are never fully adequate. Planets do not move in ellipses—elliptic trajectories are a good, but far from perfect, approximation. Matter is not made of atoms, electrons, or quarks—all these are good models (indeed, very good ones), but ones that are bound to require revision someday. A lot of conceptual difficulties could be clarified if mathematicians and theoretical physicists paid more attention to the basic distinction between model and reality, a concept familiar to biologists.

Stanislas Dehaene

13

The Assortative Mating Theory

Simon Baron-Cohen

Psychologist, Autism Research Centre, Cambridge University; Author, The Essential Difference

I've been working on the question of autism, trying to understand what characterizes autism from a psychological perspective and ultimately aiming to understand what's going on in the brain and the causes of the condition. My new theory is that it's not just a genetic condition, but it might be the result of two particular types of parents, who are both contributing genes. This might be controversially received. This is because there are a number of different theories out there—one of which is an environmental theory, such as autism being caused by vaccine damage—the MMR vaccine (the measles, mumps, and rubella combination vaccine). Another environmental theory is that autism is due to toxic levels of mercury building up in the child's brain. But the genetic theory has a lot of evidence, and what we are now testing is that if two "systemizers" have a child, this will increase the risk of the child having autism. That's it in a nutshell.

A systemizer is somebody whose style of thinking is predominantly in terms of understanding things according to rules or laws. You can think of lots of different kinds of systems: mathematical systems (algebra, computer programs); mechanical systems (computers, cars); natural systems (weather, rocks, geology); and social systems (businesses, the military).

In each case, when you systemize what you do, you try to understand the system in terms of the laws that govern the system. Economics would be an example of a system, where people are trying to predict a crash, or predict what's going to happen in terms of stock markets. They are trying to understand things according to laws or rules. The theory we are testing is that if you have a mother and a father who are both systemizers, the risk of the child having autism increases.

Systemizing is expressed in behavior, so, for example, if your hobby is playing with computers, that's the behavior that you see. But obviously such activity reflects your interests, which is what's going on in your mind, not just in your behavior. The mind of a systemizer is drawn to understand systems. Steven Pinker has a nice phrase about spiders, that spiders are just programmed to spin webs. He uses that as an analogy for the way in which a typically developing child is programmed to learn language. These programs are not 100 percent deterministic; you can intervene, you can change. There's obviously plasticity in the system. In the same way, systemizing isn't going to turn out to be 100 percent genetic. There are few if any behavioral characteristics in humans that are 100 percent genetic.

There are five steps for testing this theory. First, we need to establish whether or not systemizing runs in families. Second, we need to find out if there are any genes associated with systemizing. Third, are the parents of children with autism systemizers, defined according to their cognition? Fourth, do they both carry the genes for systemizing? Finally, when these genes combine, does this raise the risk of their child having autism?

This theory will be controversial, and it might raise anxieties. But just because it's potentially controversial doesn't mean that we shouldn't investigate it. And there are ways that you can investigate it empirically.

How would you investigate it? Well, first thing is to look at families where there's already an autistic child, and look at the parents directly. We've already conducted some of those studies, and found that whereas in the general population systemizing is more common

Simon Baron-Cohen

among males, in the case of parents of a child with autism, the mother of such a child is also very likely to be a systemizer, with male-typical interests.

One example of how we test this is to give them a task where you have to analyze a visual pattern as quickly as you can to find a component part. In the general population males are quicker at this kind of analytic task, but in the case of parents of children with autism, the mothers are just as fast as typical males. The mothers are showing a typical male profile, and that's counterintuitive since you would expect them to be showing a more typical female profile. That's just one clue that this theory is worth exploring.

A second clue is that we've looked at the rates of engineering in both fathers and the grandfathers of children with autism. Engineering is an occupation where you have to be a good systemizer—for example, understanding mechanical systems. We found that fathers of children with autism are overrepresented in the field of engineering. And what was interesting was that we found exactly that same pattern in the grandparents, too.

You start with the child with autism; he or she is the end result of this experiment of nature. And you work backward to see if there were clues in the previous generation—or previous two generations. This new theory is called "the assortative mating theory." The clue that both sides of the family are contributing similar genes is that in our study of occupations, grandfathers on the maternal and the paternal sides were both more likely to be working in the field of engineering. So the strong systemizing wasn't coming down just one side of the family. It's called assortative mating because it describes the idea that two individuals might end up in a union because of having similar characteristics. They're selecting each other on the basis of having similar characteristics.

The assortative mating theory connects with the field of sex differences—my other big area of interest. I've been trying to understand the differences between males and females. It was interesting for me to discover that there's been a sleight of hand, mostly in the States, such that the word "sex" has been replaced by the word "gender."

This has happened in a very subtle way over the last century, so that in the States, nobody talks about sex differences; they talk about gender differences. Whenever you want to refer to somebody's sex you refer to their gender. I call it a sleight of hand, because actually "sex" is the older word. Your sex is either male or female, and in biology your sex is defined by whether you have 2 X chromosomes or an X and Y chromosome. There's been a subtle shift into talking about gender, to whitewash the word "sex."

Why has this happened? Presumably because your sex is determined by your chromosomes. And in the States the ideology is that we shouldn't be determined by anything; we should be able to be anything we choose. The blank slate. Gender refers to how you think of yourself: as masculine or feminine. It's much more subjective, and is commonly believed to be culturally constructed. Italian male gender behavior is expressed differently from English male gender behavior. This gives the impression that people's gender behavior can change as they change culture, even if their biological sex is fixed.

Talking about gender is therefore much more optimistic than talking about sex. It's the rags to riches idea—you can become anything. But I've been very interested to go back to the original notion of sex, as a biological characteristic, and to ask if there are any essential differences between males and females in the mind. And to understand that if there are psychological differences, what are the biological mechanisms that give rise to these? Are they genes, are they hormones?

In our own work, we have been focused very much on fetal testosterone—the hormone that the fetus is producing in the womb—to see whether that has any effect on later behavior. We had

Simon Baron-Cohen

a perfectly good word, which was "sex." But it's become almost a profane word in the U.S. I recently wrote a journal article on sex differences in the mind. Everywhere I'd written the word "sex," the copy editor changed it to the word "gender." A systematic change had been introduced, and I asked that the original word be used. The editors asked me to give them a good reason, because they explained in the States the preferred word is gender. I had to explain, a person's gender is different to their sex. It's a distinction that seems to have got lost. It's hard to know whether it was deliberate, or whether it just happened without anybody noticing.

Back to hormones. We've been conducting laboratory studies on the amniotic fluid in the womb—the fetus is effectively swimming in this amniotic fluid. We analyze how much testosterone, the so-called male hormone, is in the amniotic fluid. It's not actually a male hormone, because both sexes produce it, it's just that males produce a lot more than females. That's because it comes from the testes. Females also produce it in the adrenal glands. And even within the boys, or within the girls, you see individual differences in how much is produced.

The question is, does this translate into anything psychological if you follow up those children? We measured the amniotic fluid testosterone, then waited until the baby was born, and then looked at the babies at twelve months old, eighteen months old, two years old. It's a longitudinal prospective study.

What we found is that the higher the baby's level of fetal testosterone, the less eye contact the child makes at twelve months old. And also the slower they are to develop language at eighteen months old. To me these are really fascinating results, because we're looking at something biological, in this case a hormone that presumably is influencing brain development to produce these quite marked differences in behavior. We always knew that girls talked earlier than boys—that there is sex difference in language development—and we also knew that there's huge variability at eighteen months: some kids have no words at all, and other kids have huge vocabularies, about 600 words. No one's really been able to explain this variability. Why should one kid be almost mute and another kid be very verbal?

People have identified some factors, such as that firstborn children talk earlier than laterborn children. Obviously there are environmental factors that are relevant. Presumably that's because firstborn children get much more attention from their parents. But over and above your birth order, it looks like hormones also explain some of the variability. We've now followed up these kids into school, they're four years old, and we're still finding that the prenatal hormone production levels are influencing behavior in middle childhood. This is just one example of why we shouldn't ignore biology in explaining differences in how the mind works.

I don't argue it's all biology. But for a long time social behavior and language development were seen as purely environmental or learned experience. These hormone studies suggest hormones are also part of the explanation. We also know from medical conditions that if, for example, for genetic reasons you have an overproduction of testosterone, this condition can change your behavior. So if you look at girls with a condition called congenital adrenal hyperplasia (CAH), where they are producing too much testosterone for genetic reasons, they look like tomboys. Their interests are very male-typical interests; they like playing with little toy cars, they like building tree houses, and they perform very quickly on spatial tests, unlike typical girls.

The evidence for my assertions comes from experiments. And in all of these areas I'm not interested in beliefs without evidence.

One experiment we conducted here in Cambridge was at the local maternity hospital. Essentially we wanted to find out whether sex differences that you observe later in life could be traced back to birth, to see if such differences are present at birth. In this experiment we looked at just over one hundred newborn babies, twenty-four hours

Simon Baron-Cohen

old, which was the youngest we could see them, and we presented each baby with a human face to look at, and then a mechanical mobile suspended above the crib. Each baby got to see both objects.

Obviously these objects are different in interesting ways, because the human face is alive, and it can express emotion, it's a natural object. The mechanical mobile is man-made, it's not alive, and obviously it doesn't have emotions. We tried to make the two objects equivalent in some important ways. One is that they were both the same size; another was that they were a similar color, in order to try and control features that might be grabbing the child's attention. But effectively what we did was film how long each baby looked at each of these two objects.

We asked the mothers not to tell us the sex of their babies, so that we could remain blind to whether this was a boy or a girl. And for the most part that was possible. Sometimes it was possible to guess that this was a boy or a girl, because there would be cards around the bed saying, "Congratulations, it's a boy." That potentially could have undermined the experiment, although we then gave the videotapes to a panel of judges to simply measure how long the baby looked at the face or the mobile. By the time the judges were looking at these videotapes they didn't have any of these potential clues to the sex of the baby, because all you could see was the eyes of the baby.

The results of the experiment were that we found more boys than girls looked longer at the mechanical mobile. And more girls than boys looked longer at the human face. Given that it was a sex difference that emerged at birth, it means that you can't attribute the difference to experience or culture. Twenty-four hours old. Now, you might say, well, they're not exactly newborn, it would have been better to get them at twenty-four minutes old—or even younger. But obviously we had to respect the wishes of the parents and the doctors to let the baby relax after the trauma of being born. And let the parents get to know their baby. So strictly speaking, it might have been one day of social experience. But nonetheless, this difference is emerging so early that it suggests it's at least partly biological.

The results were published in 2001 in a scientific journal and the experiment hasn't yet been replicated, and obviously in science what is needed is independent replication. I'll be interested in other labs to attempt to do this. As far as I know there hasn't been any attempt. This may be because it's quite hard work.

To test a hundred babies, you have to hang around hospitals waiting for babies to be born. That sounds pretty straightforward, because babies are being born every day. In a city like Cambridge there are about five new babies born a day. For some reason babies tend to be born in the middle of the night, about two or three o'clock in the morning. You have to have a very dedicated research team who are willing to wait. In Cambridge, mothers only stay in hospital for one day. Maybe one night. Then they are sent home, in order to vacate the bed for another expectant mother. In terms of your window of opportunity for testing babies, you therefore have to be there at the right time. We had two very hardworking master's students who approached mothers to ask for parental consent—maybe that was easier in a city like Cambridge, because parents know that in a university town research is going on.

The test is not invasive—the baby just has to lie on their back and look up. They were presented with each object for only one minute, because babies tend to get very restless very quickly. It's a difficult experiment to conduct, because babies spend most of their time sleeping, or feeding, or crying. You have to wait until they're not doing any of those three things. When they're awake and calm, you have a couple of minutes to present the stimuli.

The camera is well-hidden off to one side. Babies can't see very far—the depth of vision of a newborn baby is only between fifteen and twenty centimeters. So it is unlikely that the presence of the camera itself affected how the baby responded.

I was expecting the experiment to be received more controversially, because as far as I know it is one of the first demonstrations of a sex difference in the mind at birth. In fact, it was published without any fuss. It may be simply that the climate has now changed, and that people are much more willing to accept that there are sex differences in the mind, and that these might even be partly biological. If that's true, then this is good news for scientists who are interested in how the mind works.

My thesis with regard to sex differences is quite moderate, in that I do not discount environmental factors; I'm just saying, don't forget about biology. To me that sounds very moderate. But for some people in the field of gender studies, even that is too extreme. They want it to be all environment and no biology. You can understand that politically that was an important position in the 1960s, in an effort to try to change society. But is it a true description, scientifically, of what goes on? It's time to distinguish politics and science, and just look at the evidence.

14

Toxo: The Parasite that Is Manipulating Human Behavior
Robert Sapolsky

Neurobiologist, Stanford University; Author, A Primate's Memoir *and* Why Zebras Don't Get Ulcers: A Guide to Stress, Stress-Related Diseases, and Coping

In the endless sort of struggle that neurobiologists have—in terms of free will, determinism—my feeling has always been that there's not a whole lot of free will out there, and if there is, it's in the least interesting places and getting more sparse all the time. But there's a whole new realm of neuroscience which I've been thinking about, which I'm starting to do research on, that throws in another element of things going on below the surface affecting our behavior. And it's got to do with this utterly bizarre world of parasites manipulating our behavior. It turns out that this is not all that surprising. There are all sorts of parasites out there that get into some organism, and what they need to do is parasitize the organism and increase the likelihood that they, the parasite, will be fruitful and multiply, and in some cases they can manipulate the behavior of the host.

Some of these are pretty astounding. There's this barnacle that rides on the back of some crab and is able to inject estrogenic hormones into the crab if the crab is male, and at that point, the male's behavior becomes feminized. The male crab digs a hole in the sand for his eggs, except he has no eggs, but the barnacle sure does, and has just gotten this guy to build a nest for him. There are other ones where wasps parasitize caterpillars and get them to defend the wasp's nests for them. These are extraordinary examples.

The parasite my lab is beginning to focus on is one in the world of mammals, where parasites are changing mammalian behavior. It's got to do with this parasite, this protozoan called Toxoplasma. If you're ever pregnant, if you're ever around anyone who's pregnant, you know you immediately get skittish about cat feces, cat bedding, cat everything, because it could carry Toxo. And you do not want to get Toxoplasma into a fetal nervous system. It's a disaster.

The normal life cycle for Toxo is one of these amazing bits of natural history. Toxo can only reproduce sexually in the gut of a cat. It comes out in the cat feces, feces get eaten by rodents. And Toxo's evolutionary challenge at that point is to figure out how to get rodents inside cats' stomachs. Now, it could have done this in really unsubtle ways, such as cripple the rodent or some such thing. Toxo instead has developed this amazing capacity to alter innate behavior in rodents.

If you take a lab rat who is 5,000 generations into being a lab rat, since the ancestor actually ran around in the real world, and you put some cat urine in one corner of their cage, they're going to move to the other side. Completely innate, hardwired reaction to the smell of cats, the cat pheromones. But take a Toxo-infected rodent, and they're no longer afraid of the smell of cats. In fact, they become attracted to it. The most damn amazing thing you can ever see, Toxo knows how to make cat urine smell attractive to rats. And rats go and check it out and that rat is now much more likely to wind up in the cat's stomach. Toxo's circle of life completed.

This was reported by a group in the U.K. about half a dozen years ago. Not a whole lot was known about what Toxo was doing in the brain, so ever since, part of my lab has been trying to figure out the neurobiological aspects. The first thing is that it's for real. The rodents, rats, mice, really do become attracted to cat urine when they've been infected with Toxo. And you might say, okay, well, this is a rodent doing just all sorts of screwy stuff because it's got this parasite turning its brain into Swiss cheese or something. It's just nonspecific behavioral chaos. But no, these are incredibly normal animals. Their olfaction is normal, their social behavior is normal, their learning and memory is normal. All of that. It's not just a generically screwy animal.

You say, okay, well, it's not that, but Toxo seems to know how to destroy fear and anxiety circuits. But it's not that, either. Because these are rats who are still innately afraid of bright lights. They're nocturnal animals. They're afraid of big, open spaces. You can condition them to be afraid of novel things. The system works perfectly well there. Somehow Toxo can laser out this one fear pathway, this aversion to predator odors.

We started looking at this. The first thing we did was introduce Toxo into a rat, and it took about six weeks for it to migrate from its gut up into its nervous system. And at that point, we looked to see, where has it gone in the brain? It formed cysts, sort of latent, encapsulated cysts, and it wound up all over the brain. That was deeply disappointing.

But then we looked at how much winds up in different areas of the brain, and it turned out Toxo preferentially knows how to home in on the part of the brain that is all about fear and anxiety, a brain region called the amygdala. The amygdala is where you do your fear conditioning; the amygdala is what's hyperactive in people with posttraumatic stress disorder; the amygdala is all about pathways of predator aversion, and Toxo knows how to get in there.

Next, we then saw that Toxo would take the dendrites, the branch and cables that neurons have to connect to each other, and shriveled them up in the amygdala. It was disconnecting circuits. You wind up with fewer cells there. This is a parasite that is unwiring this stuff in the critical part of the brain for fear and anxiety. That's really interesting. That doesn't tell us a thing about why only its predator aversion has been knocked out, whereas fear of bright lights, et cetera, is still in there. It knows how to find that particular circuitry.

So what's going on from there? What's it doing? Because it's not just destroying this fear aversive response, it's creating something new. It's creating an attraction to the cat urine. And here is where this gets utterly bizarre. You look at circuitry in the brain, and there's a reasonably well-characterized circuit that activates neurons which become metabolically active circuits where they're talking to each other, a reasonably well-understood process that's involved in predator aversion. It involves neurons in the amygdala, the hypothalamus, and some other brain regions getting excited. This is a very well-characterized circuit.

Meanwhile, there is a well-characterized circuit that has to do with sexual attraction. And as it happens, part of this circuit courses through the amygdala, which is pretty interesting in and of itself, and then goes to different areas of the brain than the fear pathways.

When you look at normal rats, and expose them to cat urine, cat pheromones, exactly as you would expect, they have a stress response: their stress hormone levels go up, and they activate this classical fear circuitry in the brain. Now you take Toxo-infected rats, right around the time when they start liking the smell of cat urine, you expose them to cat pheromones, and you don't see the stress hormone release. What you see is that the fear circuit doesn't activate normally, and instead the sexual arousal activates some. In other words, Toxo knows how to hijack the sexual reward pathway. And you get males infected with Toxo and expose them to a lot of the cat pheromones, and their testes get bigger. Somehow, this damn parasite knows how to make cat urine smell sexually arousing to rodents, and they go and check it out. Totally amazing.

So on a certain level, that explains everything. Aha! It takes over sexual arousal circuitry. This is utterly bizarre. At this point, we don't know what the basis is of the attraction in the females. It's something we're working on.

Some extremely nice work has been done by a group at Leeds in the U.K., who are looking at the Toxo genome, and we're picking up on this collaboratively. Okay, Toxo, it's a protozoan parasite. Toxo and mammals had a common ancestor, and the last time they did was, God knows, billions of years ago. And you look in the Toxo genome, and it's got two versions of the gene called tyrosine hydroxylase. And if you were a neurochemistry type, you would be leaping up in shock and excitement at this point.

Tyrosine hydroxylase is the critical enzyme for making dopamine: the neurotransmitter in the brain that's all about reward and anticipation of reward. Cocaine works on the dopamine system; all sorts of other euphoriants do. Dopamine is about pleasure, attraction, and anticipation. And the Toxo genome has the mammalian gene for making the stuff. It's got a little tail on the gene that targets, specifies, that when this is turned into the actual enzyme, it gets secreted out of the Toxo and into neurons. This parasite doesn't need to learn how to make neurons act as if they are pleasurably anticipatory; it takes over the brain chemistry of it all on its own.

Again that issue of specificity comes up. Look at closely related parasites to Toxo: Do they have this gene? Absolutely not. Now look at the Toxo genome and look at genes related to other brain messengers. Serotonin, acetylcholine, norepinephrine, and so on, and you go through every single gene you can think of. Zero. Toxo doesn't have them; Toxo's got this one gene which allows it to just plug into the whole world of mammalian reward systems. And at this point, that's what we know. It is utterly cool.

Of course, at this point, you say well, what about other species? What does Toxo do to humans? And there's some interesting stuff there that's reminiscent of what's going on in rodents. Clinical dogma is you first get a Toxo infection. If you're pregnant, it gets into the fetal nervous system, a huge disaster. Otherwise, if you get a Toxo infection, it has phases of inflammation, but eventually it goes into this latent asymptomatic stage, which is when these cysts form in the brain. Which is, in a rat, when it stops being anything boring like asymptomatic, and when the behavior starts occurring. Interestingly, that's when the parasite starts making this tyrosine hydroxylase.

So what about humans? A small literature is coming out now reporting neuropsychological testing on men who are Toxo-infected, showing that they get a little bit impulsive. Women less so, and this may have some parallels perhaps with this whole testosterone aspect of the story that we're seeing. And then the truly astonishing thing: Two different groups independently have reported that people who are Toxo-infected have three to four times the likelihood of being killed in car accidents involving reckless speeding.

In other words, you take a Toxo-infected rat and it does some dumb-ass thing that it should be innately skittish about, like going right up to cat smells. Maybe you take a Toxo-infected human and they start having a proclivity toward doing dumb-ass things that we should be innately averse to, like having your body hurdle through space at high g-forces. Maybe this is the same neurobiology. This is not to say that Toxo has evolved the need to get humans into cat stomachs. It's just sheer convergence. It's the same nuts-and-bolts neurobiology in us and in a rodent, and does the same thing.

On a certain level, this is a protozoan parasite that knows more about the neurobiology of anxiety and fear than 25,000 neuroscientists standing on each other's shoulders, and this is not a rare pattern. Look at the rabies virus; rabies knows more about aggression than we neuroscientists do. It knows how to make you rabid. It knows how to make you want to bite someone, and that saliva of yours contains rabies virus particles, passed on to another person.

The Toxo story is, for me, completely new terrain—totally cool, interesting stuff, just in terms of this individual problem. And maybe it's got something to do with treatments for phobias down the line, or whatever it is to make it seem like anything more than just the coolest gee-whiz thing possible. But no doubt it's also a tip of the iceberg of God knows what other parasitic stuff is going on out there. Even in the larger sense, God knows what other unseen realms of biology make our behavior far less autonomous than lots of folks would like to think.

With regard to parasite infections like Toxo in humans, there is a big prevalence in certain parts of the world. There's a higher prevalence in the tropics, where typically more than 50 percent of people are infected. Lower rates in more temperate zones for reasons that I do not understand and do not choose to speculate on. France has really high rates of Toxo infection. In much of the developing world, it's bare feet, absorbing it through soil, where cats may have been. It's food that may not have been washed sufficiently and absorption through hands. It's the usual story that people in the developing world are more subject to all sorts of infectious stuff.

A few years ago, I sat down with a couple of the Toxo docs over in our hospital who do the Toxo testing in the ob-gyn clinics. And they hadn't heard about this behavioral story, and I'm going on about how cool and unexpected it is. And suddenly, one of them jumps up, flooded with forty-year-old memories, and says, "I just remembered back when I was a resident, I was doing a surgical transplant rotation. And there was an older surgeon, who said, if you ever get organs from a motorcycle accident death, check the organs for Toxo. I don't know why, but you find a lot of Toxo." And you could see this guy was having a rush of nostalgic memories from back when he was twenty-five, and all because he was being told this weird factoid . . . ooh, people who die in motorcycle accidents seem to have high rates of Toxo. Utterly bizarre.

What is the bottom line on this? Well, it depends; if you want to overcome some of your inhibitions, Toxo might be a very good thing to have in your system. Not surprisingly, ever since we started studying Toxo in my lab, every lab meeting we sit around speculating about which people in the lab are Toxo-infected, and that might have something to do with one's level of recklessness. Who knows? It's very interesting stuff, though.

You want to know something utterly terrifying? Here's something terrifying and not surprising. Folks who know about Toxo and its affect on behavior are in the U.S. military. They're interested in Toxo. They're officially intrigued. And I would think they would be intrigued, studying a parasite that makes mammals perhaps do things that everything in their fiber normally tells them not to because it's dangerous and ridiculous and stupid and don't do it. But suddenly with this parasite on board, the mammal is a little bit more likely to go and do it. Who knows? But they are aware of Toxo.

There are two groups that collaborate in Toxo research. One is Joanne Webster, who was at Oxford at the time that she first saw this behavioral phenomenon. And I believe she's now at University College London. And the other is Glenn McConkey at University of Leeds. And they're on this. She's more of a behaviorist; he's more of an enzyme biochemist guy. We're doing the neurobiology end of it. We're all talking lots.

There's a long-standing literature that absolutely shows there's a statistical link between Toxo infection and schizophrenia. It's not a big link, but it's solidly there. Schizophrenics have higher than expected rates of having been infected with Toxo, and that's not particularly the case for other related parasites. Links between schizophrenia and mothers who had house cats during pregnancy. There's a whole literature on that. So where does this fit in?

Two really interesting things. Back to dopamine and the tyrosine hydroxylase gene that Toxo somehow ripped off from mammals, which allows it to make more dopamine. Dopamine levels are too

Robert Sapolsky

high in schizophrenia. That's the leading suggestion of what schizophrenia is about neurochemically. You take Toxo-infected rodents and their brains have elevated levels of dopamine. Final deal is, and this came from Webster's group, you take a rat who's been Toxo-infected and is now at the state where it would find cat urine to be attractive, and you give it drugs that block dopamine receptors, the drugs that are used to treat schizophrenics, and it stops being attracted to the cat urine. There is some schizophrenia connection here with this.

Any time Toxo's picked up in the media, and this schizophrenia angle is brought in, the irresistible angle is the generic crazy cat lady, you know, living in the apartment with forty-three cats and their detritus. And that's an irresistible one in terms of Toxo psychiatric status: cats. But God knows what stuff is lurking there.

15

Amazing Babies

Alison Gopnik

Psychologist, University of California, Berkeley; Author, The Philosophical Baby

The biggest question for me is, "How is it possible for children, young human beings, to learn as much as they do as quickly and as effectively as they do?" We've known for a long time that human children are the best learning machines in the universe. But it has always been like the mystery of the hummingbirds. We know that they fly, but we don't know how they can possibly do it. We could say that babies learn, but we didn't know how.

But now there's this really exciting confluence of work in artificial intelligence and machine learning, neuroscience, and in developmental psychology, all trying to tackle this question about how children could possibly learn as much as they do.

What's happened is that there are more and more really interesting models coming out of AI and machine learning. Computer scientists and philosophers are starting to understand how scientists or machines or brains could actually do something that looks like powerful inductive learning. The project we've been working on for the past ten years or so is to ask whether children and even young babies implicitly use some of those same really powerful inductive learning techniques.

It's been very exciting because, on the one hand, it helps to explain the thing that's been puzzling developmental psychologists since Piaget. Every week we discover some new amazing thing about what babies and young children know that we didn't realize before. And then we discover some other amazing thing about what they don't yet know. So we've charted a series of changes in children's knowledge—we know a great deal about what children know when. But the great mystery is how could they possibly learn it? What computations are they performing? And we're starting to answer that.

It's also been illuminating because the developmentalists can help the AI people do a sort of reverse engineering. When you realize that human babies and children are these phenomenal learners, you can ask, okay, what would happen if we actually used what we know about children to help program a computer?

The research starts out from an empirical question and a practical question, How do children learn so much? How could we design computers that learn? But then it turns out that there's a big grand philosophical question behind it. How do any of us learn as much as we do about the world? All we've got are these little vibrations of air in our eardrums and photons hitting the back of our retina. And yet human beings know about objects and people, not to mention quarks and electrons. How do we ever get there? How could our brains, evolved in the Pleistocene, get us from the photons hitting our retinas to quarks and electrons? That's the big, grand philosophical question of knowledge.

Understanding how we learn as children actually ends up providing at least the beginning of an answer to this much bigger philosophical question. The philosophical convergence, which also has a nice moral quality, is that these very, very high-prestige learning systems like scientists and fancy computers at Microsoft turn out to be doing similar things to the very, very low-prestige babies and toddlers and preschoolers. Small children aren't the sort of people that philosophers and psychologists and scientists have been paying much attention to over the last 2,000 years. But just looking at these babies and little kids running around turns out to be really informative about deep philosophical questions.

Alison Gopnik

For example, it turns out that babies and very young children already are doing statistical analyses of data, which is not something that we knew about until the last ten years. This is a really very, very new set of findings. Jenny Saffran, Elissa Newport, and Dick Aslin at Rochester started it off when they discovered that infants could detect statistical patterns in nonsense syllables. Now every month there's a new study that shows that babies and young children compute conditional probabilities, that they do Bayesian reasoning, that they can take a random sample and understand the relationship between that sample and the population that it's drawn from. And children don't just detect statistical patterns, they use them to infer the causal structure of the world. They do it in much the same way that sophisticated computers do. Or for that matter, they do it in the same way that every scientist does who looks at a pattern of statistics and doesn't just say oh, that's the data pattern, but can then say oh, and that data pattern tells us that the world must be this particular way.

How could we actually ask babies and young children to tell us whether they understand statistics? We know that when we even ask adults to actually explicitly solve a probability problem, they collapse. How could we ask little kids to do it?

The way that we started out was that we built a machine we called the blicket detector. The blicket detector is a little machine that lights up and plays music when you put certain things on it but not others. We can actually control the information that the child gets about the statistics of this machine. We put all sorts of different things on it. Sometimes the box lights up, sometimes it doesn't, sometimes it plays music, sometimes it doesn't. And then we can ask the child things like what would happen if I took the yellow block off? Or which block will make it go best? And we can design it so that, for example, one block makes it go two out of eight times, and one block makes it go two out of three times.

Four-year-olds, who can't add yet, say that the block that makes it go two out of three times is a more powerful block than the one that makes it go two out of eight times. That's an example of the kind of implicit statistics that even two- and three- and four-year-olds are using when they're trying to just figure out something about how this particular machine goes. And we've used similar experiments to show that children can use Bayesian reasoning, infer complex causal structure, and even infer hidden, invisible causal variables.

With even younger babies, Fei Xu showed that nine-month-olds were already paying attention to the statistics of their environment. She would show the baby a box of mostly red Ping-Pong balls, 80 percent red, 20 percent white. And then a screen would come up in front of the Ping-Pong balls and someone would take out a number of Ping-Pong balls from that box. They would pick out five red Ping-Pong balls or else pick out five white Ping-Pong balls. Well, of course, neither of those events is impossible. But picking out five white Ping-Pong balls from an 80 percent red box is much less likely. And even nine-month-olds will look longer when they see the white Ping-Pong balls coming from the mostly red box than when they see the red Ping-Pong balls coming from the mostly red box.

Fei did a beautiful control condition. Exactly the same thing happens, except now instead of taking the balls from the box, the experimenter takes the balls from her pocket. When the experimenter takes the balls from her pocket, the baby doesn't know what the population is that the experimenter is sampling. And in that case, the babies don't show any preference for the all-red versus all-white sample. The babies really seem to have an idea that some random samples from a population are more probable, and some random samples from a population are less probable.

The important thing is not just that they know this, which is amazing, but that once they know this, then they can use that as a foundation for making all sorts of other inferences. Fei and Henry Wellman and one of my ex-students, Tamar Kushnir, have been doing studies where you show babies the unrepresentative sample . . . someone picks out five white balls from a mostly red box. And now there are red balls and white balls on the table and the experimenter puts her hand out and says, "Give me some."

Well, if the sample wasn't representative, then you think, well, okay, why would she have done that? She must like the white balls. And, in fact, when the sample's not representative, the babies give her the white balls. In other words, not only do the babies recognize whether this is a random sample or not, but when it isn't random, they say oh, this isn't just a random sample, there must be something else going on. And by the time they're eighteen months old, they seem to think oh, the thing that's going on is that she would rather have white balls than red balls.

Not only does this show that babies are amazing, but it actually gives the babies a mechanism for learning all sorts of new things about the world. We can't ask these kids explicitly about probability and statistics, because they don't yet understand that two plus two equals four. But we can look at what they actually do and use that as a way of figuring out what's going on in their minds. These abilities provide a framework by which the babies can learn all sorts of new things that they're not innately programmed to know. And that helps to explain how all humans can learn so much, since we're all only babies who have been around for a while.

Another thing that it turns out that kids are doing is that they're experimenting. You see this just in their everyday play. They are going out into the world and picking up a toy and pressing the buttons and pulling the strings on it. It looks like random play, but when you look more carefully, it turns out that that apparently random play is actually a set of quite carefully done experiments that let children figure out how it is that that toy works. Laura Schulz at MIT has done a beautiful set of studies on this.

The most important thing for children to figure out is us, other human beings. We can show that when we interact with babies they recognize the contingencies between what we do and they do. Those are the statistics of human love. I smile at you and you smile at me. And children also experiment with people, trying to figure out what the other person is going to do and feel and think. If you think of them as little psychologists, we're the lab rats.

The problem of learning is actually in Turing's original paper that is the foundation of cognitive science. The classic Turing problem is, "Could you get a computer to be so sophisticated that you couldn't tell the difference between that computer and a person?" But Turing said that there was an even more profound problem, a more profound Turing test. Could you get a computer, give it the kind of data that every human being gets as a child, and have it learn the kinds of things that a child can learn?

The way that Chomsky solved that problem was to say: Oh, well, we don't actually learn very much. What happens is that it's all there innately. That's a philosophical answer that has a long tradition going back to Plato and Descartes and so forth. That set the tone for the first period of the cognitive revolution. And that was reinforced when developmentalists like Andrew Meltzoff, Liz Spelke, and Renee Baillargeon began finding that babies knew much more than we thought.

Part of the reason why innateness seemed convincing is because the traditional views of learning have been very narrow, like Skinnerian reinforcement or association. Some cognitive scientists, particularly connectionists and neural network theorists, tried to argue that these mechanisms could explain how children learn but it wasn't convincing. Children's knowledge seemed too abstract and coherent, too far removed from the data, to be learned by association. And, of course, Piaget rejected both these alternatives and talked about "constructivism," but that wasn't much more than a name.

Then about twenty years ago, a number of developmentalists working in the Piagetian tradition, including me and Meltzoff and

Susan Carey, Henry Wellman, and Susan Gelman, started developing the idea that I call the "theory theory." That's the idea that what babies and children are doing is very much like scientific induction and theory change.

The problem with that was that when we went to talk to the philosophers of science and we said, "Okay, how is it that scientists can solve these problems of induction and learn as much as they do about the world?" they said, "We have no idea, go ask psychologists." Seeing that what the kids were doing was like what scientists were doing was sort of helpful, but it wasn't a real cognitive science answer.

About fifteen years ago, quite independently, a bunch of philosophers of science at Carnegie Mellon, Clark Glymour and his colleagues, and a bunch of computer scientists at UCLA, Judea Pearl and his colleagues, converged on some similar ideas. They independently developed these Bayesian causal graphical models. The models provide a graphical representation of how the world works and then systematically map that representation onto patterns of probability. That was a great formal computational advance.

Once you've got that kind of formal computational system, then you can start designing computers that actually use that system to learn about the causal structure of the world. But you can also start asking, well, do people do the same thing? Clark Glymour and I talked about this for a long time. He would say oh, we're actually starting to understand something about how you can solve inductive problems. I'd say gee, that sounds a lot like what babies are doing. And he'd say no, no, come on, they're just babies, they couldn't be doing that.

What we started doing empirically about ten years ago is to actually test the idea that children might be using these computational procedures. My lab was the first to do it, but now there is a whole set of great young cognitive scientists working on these ideas. Josh Tenenbaum at MIT and Tom Griffiths at Berkeley have worked on the computational side. On the developmental side Fei Xu, who is now at Berkeley, Laura Schulz at MIT, and David Sobel at Brown, among others, have been working on empirical experiments with children. We've had this convergence of philosophers and computer scientists on the one hand, and empirical developmental psychologists on the other hand, and they've been putting these ideas together. It's interesting that the two centers of this work, along with Rochester, have been MIT, the traditional locus of "East Coast" nativism, and Berkeley, the traditional locus of "West Coast" empiricism. The new ideas really cross the traditional divide between those two approaches.

A lot of the ideas are part of what's really a kind of Bayesian revolution that's been happening across cognitive science, in vision science, in neuroscience and cognitive psychology and now in developmental psychology. Ideas about Bayesian inference that originally came from the philosophy of science have started to become more and more powerful and influential in cognitive science in general.

Whenever you get a new set of tools unexpected insights pop up. And, surprisingly enough, thinking in this formal computational nerdy way actually gives us new insights into the value of imagination. This all started by thinking about babies and children as being like little scientists, right? We could actually show that children would develop theories and change them in the way that scientists do. Our picture was . . . there's this universe, there's this world that's out there. How do we figure out how that world works?

What I've begun to realize is that there's actually more going on than that. One of the things that makes these causal representations so powerful and useful in AI is that not only do they let you make predictions about the world, but they let you construct counterfactuals. And counterfactuals don't just say what the world is like now. They say here's the way the world could be, other than the way it is now. One of the great insights that Glymour and Pearl had was that, formally, constructing these counterfactual claims was quite different

Alison Gopnik

from just making predictions. And causal graphical representations and Bayesian reasoning are a very good combination because you're not just talking about what's here and now, you're saying . . . here's a possibility, and let me go and test this possibility.

If you think about that from the perspective of human evolution, our great capacity is not just that we learn about the world. The thing that really makes us distinctive is that we can imagine other ways that the world could be. That's really where our enormous evolutionary juice comes from. We understand the world, but that also lets us imagine other ways the world could be, and actually make those other worlds come true. That's what innovation, technology, and science are all about.

Think about everything that's in this room right now, there's a right-angle desk and electric light and computers and windowpanes. Every single thing in this room is imaginary from the perspective of the hunter-gatherer. We live in imaginary worlds.

When you think that way, a lot of other things about babies and young children start to make more sense. We know, for instance, that young children have these incredible, vivid, wild imaginations. They live 24/7 in these crazy pretend worlds. They have a zillion different imaginary friends. They turn themselves into ninjas and mermaids. Nobody's really thought about that as having very much to do with real hard-nosed cognitive psychology. But once you start realizing that the reason why we want to build theories about the world is so that we can imagine other ways the world can be, you could say that not only are these young children the best learners in the world, but they're also the most creative imaginers in the world. That's what they're doing in their pretend play.

About ten years ago psychologists like Paul Harris and Marjorie Taylor started to show that children aren't confused about fantasy and imagination and reality, which is what psychologists from Freud to Piaget had thought before. They know the difference between imagination and reality really well. It's just they'd rather live in imaginary worlds than in real ones. Who could blame them? In that respect, again, they're a lot like scientists and technologists and innovators.

One of the other really unexpected outcomes of thinking about babies and children in this new way is that you start thinking about consciousness differently. Now of course there's always been this big question . . . the capital C question of consciousness. How can a brain have experiences? I'm skeptical about whether we're ever going to get a single answer to the big capital C question. But there are lots of very specific things to say about how particular kinds of consciousness are connected to particular kinds of functional or neural processes.

Edge asked a while ago in the World Question Center, what is something you believe but can't prove? And I thought well, I believe that babies are actually not just conscious but more conscious than we are. But of course that's not something that I could ever prove. Now, having thought about it and researched it for a while, I feel that I can not quite prove it, but at least I can make a pretty good empirical case for the idea that babies are in some ways more conscious, and certainly differently conscious, than we are.

For a long time, developmental psychologists like me had said, well, babies can do all these fantastic amazing things, but they're all unconscious and implicit. A part of me was always skeptical about that, though, just intuitively, having spent so much time with babies. You sit opposite a seven-month-old, and you watch their eyes and you look at their face and you see that wide-eyed expression and you say, goddamn it, of course she's conscious, she's paying attention.

We know a lot about the neuroscience of attention. When we pay attention to something as adults, we're more open to information about that thing, but the other parts of our brain get inhibited. The metaphor psychologists always use is that it's like a spotlight. It's as if what happens when you pay attention is that you shine a light on one

Alison Gopnik

particular part of the world, make that little part of your brain available for information processing, change what you think, and then leave all the rest of it alone.

When you look at both the physiology and the neurology of attention in babies, what you see is that instead of having this narrow focused top-down kind of attention, babies are open to all the things that are going on around them in the world. Their attention isn't driven by what they're paying attention to. It's driven by how information-rich the world is around them. When you look at their brains, instead of just, as it were, squirting a little bit of neurotransmitter on the part of their brain that they want to learn, their whole brain is soaked in those neurotransmitters.

The thing that babies are really bad at is inhibition, so we say that babies are bad at paying attention. What we really mean is that they're bad at not paying attention. What we're great at as adults is not paying attention to all the distractions around us, and just paying attention to one thing at a time. Babies are really bad at that. But the result is that their consciousness is like a lantern instead of being like a spotlight.

They're open to all of the experience that's going on around them.

There are certain kinds of states that we're in as adults, like when we go to a new city for the first time, where we recapture that baby information processing. When we do that, we feel as if our consciousness has expanded. We have more vivid memories of the three days in Beijing than we do of all the rest of the months that we spend as walking, talking, teaching, meeting-attending zombies. So that we can actually say something about what babies' consciousness is like, and that might tell us some important things about what consciousness itself is like.

I come from a big, close family. Six children. It was a somewhat lunatic artistic intellectual family of the 1950s and 1960s, back in the golden days of postwar Jewish life. I had this wonderful rich, intellectual and artistic childhood. But I was also the oldest sister of six children, which meant that I was spending a lot of time with babies and young children.

I had the first of my own three babies when I was twenty-three. There's really only been about five minutes in my entire life when I haven't had babies and children around. I always thought from the very beginning that they were the most interesting people there could possibly be. I can remember being in a state of mild indignation, which I've managed to keep up for the rest of my life, about the fact that other people treated babies and children contemptuously or dismissively or neglectfully.

At the same time, from the time I was very young, I knew that I wanted to be a philosopher. I wanted to actually answer, or at least ask, big, deep questions about the world, and I wanted to spend my life talking and arguing. And, in fact, that's what I did as an undergraduate. I was an absolutely straight down the line honors philosophy student as an undergraduate at McGill, president of the Philosophy Students Association, etc. I went to Oxford partly because I wanted to do both philosophy and psychology.

But what kept happening to me was that I asked these philosophical questions, and I'd say, well, you know, you could find out. You want to know where language comes from? You could go and look at children, and you could find out how children learn language. Or you want to find out how we understand about the world? You could look at children and find out how they, that is we, come to understand about the world. You want to understand how we come to be moral human beings? You could look at what happens to moral intuition in children. And every time I did that, back in those bad old days, the philosophers around me would look as if I had just eaten my peas with a knife. One of the Oxford philosophers said to me after one of these conversations, "Well, you know, one's seen children about, of course. But one would never actually talk to them." And that wasn't

Alison Gopnik

atypical of the attitude of philosophy toward children and childhood back then.

I still think of myself as being fundamentally a philosopher; I'm an affiliate of the Philosophy Department at Berkeley. I give talks at the American Philosophical Association and publish philosophical papers. It's coincidental that the technique I use to answer those philosophical questions is to look at children and think about children. And I'm not alone in this. Of course, there are still philosophers out there who believe that philosophy doesn't need to look beyond the armchair. But many of the most influential thinkers in philosophy of mind understand the importance of empirical studies of development.

In fact, largely because of Piaget, cognitive development has always been the most philosophical branch of psychology. That's true if you look not just at the work that I do, but the work that people like Andrew Meltzoff or Henry Wellman or Susan Carey or Elizabeth Spelke do, or certainly what Piaget did himself. Piaget also thought of himself as a philosopher who was answering philosophical questions by looking at children.

Thinking about development also changes the way we think about evolution. The traditional picture of evolutionary psychology is that our brains evolved in the Pleistocene, and we have these special purpose modules or innate devices for organizing the world. They're all there in our genetic code, and then they just unfold maturationally. That sort of evolutionary psychology picture doesn't fit very well with what most developmental psychologists see when they actually study children.

When you actually study children, you certainly do see a lot of innate structure. But you also see this capacity for learning and transforming and changing what you think about the world and for imagining other ways that the world could be. In fact, one really crucial evolutionary fact about us is that we have this very, very extended childhood. We have a much longer period of immaturity than any other species does. That's a fundamental evolutionary fact about us, and on the surface a puzzling one. Why make babies so helpless for so long? And why do we have to invest so much time and energy, literally, just to keep them alive?

Well, when you look across lots and lots of different species, birds and rodents and all sorts of critters, you see that a long period of immaturity is correlated with a high degree of flexibility, intelligence, and learning. Look at crows and chickens, for example. Crows get on the cover of *Science* using tools, and chickens end up in the soup pot, right? And crows have a much longer period of immaturity, a much longer period of dependence than chickens.

If you have a strategy of having these very finely shaped innate modules just designed for a particular evolutionary niche, it makes sense to have those in place from the time you're born. But you might have a more powerful strategy. You might not be very well designed for any particular niche, but instead be able to learn about all the different environments in which you can find yourself, including being able to imagine new environments and create them. That's the human strategy.

But that strategy has one big disadvantage, which is that while you're doing all that learning, you are going to be helpless. You're better off being able to consider, for example, should I attack this mastodon with this kind of tool or that kind of tool? But you don't want to be sitting and considering those possibilities when the mastodon is coming at you.

The way that evolution seems to have solved that problem is to have this kind of cognitive division of labor, so the babies and kids are really the R&D department of the human species. They're the ones that get to do the blue-sky learning, imagining, thinking. And the adults are production and marketing. We can not only function effectively but we can continue to function in all these amazing new

Alison Gopnik

environments, totally unlike the environment in which we evolved. And we can do so just because we have this protected period when we're children and babies in which we can do all of the learning and imagining. There's really a kind of metamorphosis. It's like the difference between a caterpillar and a butterfly, except it's more like the babies are the butterflies that get to flitter around and explore, and we're the caterpillars who are just humping along on our narrow adult path.

Thinking about development not only changes the way you think about learning, but it changes the way that you think about evolution. And again, it's this morally appealing reversal, which you're seeing in a lot of different areas of psychology now. Instead of just focusing on human beings as the competitive hunters and warriors, people are starting to recognize that our capacities for caregiving are also, and in many respects, even more fundamental in shaping what our human nature is like.

16

Signatures of Consciousness

Stanislas Dehaene

Neuroscientist, Collège de France, Paris; Author, The Number Sense *and* Reading in the Brain

I want to start by describing a small object. It's not extraordinary, but I sort of like it. Of course, it's a brain. In fact, it is my brain, exactly as it was ten years ago. This is not its real size, of course. It's smaller than life and is only made of plaster. It is one of the first 3-D printouts that we made from a rapid prototyping machine, one of those 3-D printers that can take the computer constructions we build from our MRI scans of the brain, and turn them into real-life 3-D objects.

This is just brain anatomy. But I'm using it to ask this question: Can this biological object, the human brain, understand itself? In the ten years that have elapsed since my brain was printed, we've made a good bit of progress in understanding some of its parts. In the lab, for instance, we've been working on an area in the parietal lobe which is related to number processing, and we've also worked on this left occipito-temporal region that we call the visual word form area, and which is related to our knowledge of spelling and reading, and whose activation develops when we learn to read.

Since this is *Edge*, the idea is not to talk about what exists and has already been published, but rather, to present new developments. So I would like to tell you about a research project that we've been working on for almost ten years and which is now running at full speed—which is trying to address the much-debated issue of the biological mechanisms of consciousness.

Neuroscientists used to wait until they were in their sixties or seventies before they dared to raise the topic of consciousness. But I now think that the domain is ripe. Today we can work on real data, rather than talk about the issue in philosophical terms.

In the past, the major problem was that people barely looked at the brain and tried to generate theories of consciousness from the top, based solely on their intuitions. Excellent physicists, for instance, tried to tell us that the brain is a quantum computer, and that consciousness will only be understood once we understand quantum computing and quantum gravity. Well, we can discuss that later, but as far as I can see, it's completely irrelevant to understanding consciousness in the brain. One of the reasons is that the temperature at which the brain operates is incompatible with quantum computing. Another is that my colleagues and I have entered an MRI scanner on a number of occasions, and have probably changed our quantum state in doing so, but as far as I can judge, this experience didn't change anything related to our consciousness.

Quantum physics is just an example. There has been an enormous wealth of theories about consciousness, but I think that very few are valid or even useful. There is, of course, the dualist notion that we need a special stuff for consciousness and that it cannot be reduced to brain matter. Obviously, this is not going to be my perspective. There is also the idea that every cell contains a little consciousness, and that if you add it all together, you arrive at an increasing amount of consciousness—again, this is also not at all my idea, as you will see.

We could go on and on, because people have proposed so many strange ideas about consciousness—but the question is, how to move forward with actual experiments. We've now done quite a few neuroimaging studies where we manage to contrast conscious and nonconscious processing, and we've tried to produce a theory from the results—so I would like to tell you about both of these aspects.

Stanislas Dehaene

Finally, at the end, I'll say a few words about perspectives for the future. One of the main motivations for my research is to eventually be in a position to apply it to patients who have suffered brain lesions and appear to have lost the ability to entertain a flow of conscious states. In such patients, the problem of consciousness is particularly apparent and vital—literally a matter of life or death. They can be in coma, in vegetative states, or in the so-called minimally conscious state. In some cases we don't even know if they are conscious or not. They could just be locked in—fully conscious, but unable to communicate, a frightening state vividly described by Jean-Dominique Bauby in *The Diving Bell and the Butterfly*. It is with such patients in mind that we must address the problem of consciousness. In the end, it's an extremely practical problem. Theories are fine, but we have to find ways that will allow us to get back to the clinic.

How to Experiment With Conscious States

So how do we experiment with consciousness? For a long time, I thought that there was no point in asking this question, because we simply could not address it. I was trained in a tradition, widely shared in the neuroscience and cognitive psychology communities, according to which you just cannot ask this question. Consciousness, for them, is not a problem that can be addressed. But, I now think this is wrong. After reading *A Cognitive Theory of Consciousness*, a book by Bernard Baars, I came to realize that the problem can be reduced to questions that are simple enough that you can test them in the lab.

I want to say right from the start that this means that we have to simplify the problem. Consciousness is a word with many different meanings, and I am not going to talk about all of these meanings. My research addresses only the most simple meaning of consciousness. Some people, when they talk about consciousness, think that we can only move forward if we gain an understanding of "the self"—the sense of being I or Me. But I am not going to talk about that. There is also a notion of consciousness as a "higher order" or "reflexive" state of mind—when I know that I know. Again, this meaning of the term remains very difficult to address experimentally. We have a few ideas on this, but it's not what I want to talk about this evening.

Tonight I only want to talk about the simpler and addressable problem of what we call "access to consciousness." At all times the brain is constantly bombarded with stimulation—and yet, we are only conscious of a very small part of it. In this room, for instance, it's absolutely obvious. We are conscious of one item here and another item there, say the presence of John behind the camera, or of some of the bottles on this table. You may not have noticed that there is a red label on the bottles. Although this information has been present on your retina since the beginning of my talk, it's pretty obvious that you are only now actually paying attention to it.

In brief, there is a basic distinction between all the stimuli that enter the nervous system, and the much smaller set of stimuli that actually make it into our conscious awareness. That is the simple distinction that we are trying to capture in our experiments. The first key insight, which is largely due to Francis Crick and Christof Koch, is that we must begin with the much simpler problem of understanding the mechanisms of access to consciousness for simple visual stimuli before we can attack the issue of consciousness and the brain.

The second key insight is that we can design minimal experimental contrasts to address this question. And by minimal contrasts, I mean that we can design experimental situations in which, by changing a very small element in the experiment, we turn something that is not conscious into something that is conscious. Patrick Cavanagh, who is here in the room, designed a large number of such illusions and stimulation paradigms. Tonight, I'll just give you one example, the so-called subliminal images that we've studied a lot.

If you flash words on a screen for a period of roughly 30 milliseconds, you see them perfectly. The short duration, in itself, is not a problem. What matters is that there is enough energy in the stimulus for you to see it. If, however, just after the word you present another string of letters at the same location, you only see the string, not the word. This surprising invisibility occurs in a range of delays between the word and the consonant string (what we call the mask) that are on the order of 50 milliseconds. If the delay is shorter than 50 milliseconds, you do not see the hidden word. It's a well-known perceptual phenomenon called masking.

Now, if you lengthen the delay a little, you see the word each time. There is even a specific delay where the subjects can see the stimulus half of the time. So now you are in business, because you have an experimental manipulation that you can reproduce in the lab, and that systematically creates a change in consciousness.

Subjective Reports Versus Objective Performance

Of course, to define our experimental conditions, we are obliged to rely on the viewer's subjective judgments. This is a very important point—we do not rely simply on a change in the stimulus. What counts is a change in stimulus that the subject claims makes his perception switch from nonconscious to conscious. Here we are touching on a difficult point—how do we define whether a subject is conscious or not? In past research, many people have been very reluctant to use this sort of subjective report. Some have argued that it is very difficult or even impossible to do science based on subjective reports. But my point of view, and I share this with many others, is that subjective reports define the science of consciousness. That's the very object we have to study—when are subjects able to report something about their conscious experience, and when are they not.

There can be other definitions. Some researchers have tried to propose an objective definition of consciousness. For instance, they have argued that, if the subject is able, say, to classify words as being animals or not, or as being words in the lexicon or not, then they are necessarily conscious. Unfortunately, however, sticking to that type of definition, based solely on an objective criterion, has been very difficult. We have repeatedly found that even when subjects claim to be unaware of the word and report that they cannot see any word at all, they still do better than chance on this type of classification task. So the problem with this approach is that we need to decide which tasks are just manifestations of subliminal or unconscious processing, and which are manifestations of access to consciousness.

In the end, however, the opposition between objective and subjective criteria for consciousness is exaggerated. The problem is not that difficult because, in fact, both phenomena often covary in a very regular fashion, at least on a broad scale. For instance, when I vary the delay between my word and my mask, what I find is that the subjects' performance suddenly increases, at the very point where they become able to report the word's presence and identity. This is an experimental observation that is fairly simple, but I think important: When subjects are able to report seeing the word, they simultaneously find many other tasks feasible with a greater rate of success.

It's not that subjects cannot react below this consciousness threshold. There is clearly subliminal processing on many tasks, but as one crosses the threshold for consciousness, there are a number of tasks that suddenly become feasible. These include the task of subjective report. Our research program consists in characterizing the major transition—from below the threshold for consciousness to above the threshold for consciousness.

I'm only giving you one example of masking, but there are now many experimental paradigms that can be used to make the same stimulus go in or out of consciousness using minimal manipulation, occasionally no manipulation at all. Sometimes the brain does the switching itself, such as, for instance, in binocular rivalry, where the two eyes see two different images but the brain only allows them to see one or the other, never both at the same time. Here the brain does the switching.

Although I'm only going to talk about the masking paradigm, because that is what we have focused on in my lab, I hope that you now understand why the experimental study of consciousness has developed into such a fast-growing field and why so many people are convinced that it is possible to experiment in this way, through the creation of minimal contrasts between conscious and nonconscious brain states.

Signatures of the Transition from Nonconscious to Conscious

Of course, we need to combine our capacity to create such minimal contrasts with methods that allow us to see the living, active brain. We are very far from having seen the end of the brain imaging revolution—this is only the beginning. Although it is hard to remember what it was like before brain imaging existed, you have to realize how amazing it is that we can now see through the skull as if it were transparent. Not only do we see the anatomy of the brain, but also how its different parts are activated, and with other techniques, the temporal dynamics with which these activations unfold. Typically, functional magnetic resonance imaging (fMRI) only lets you see the static pattern of activation on a scale of one or two seconds. With other techniques, such as electro- or magneto-encephalography, however, you can really follow in time, millisecond by millisecond, how that activation progresses from one site to the other.

What do we see when we do these experiments? The first thing that we discovered is that even when you cannot see a word or a picture, because it is presented in a subliminal condition, it does not mean that your cortex is not processing it. Some people initially thought that subliminal processing meant subcortical processing—processing that is not done in the cortex. It's of course completely false and we've known this for a while now. We can see a lot of cortical activation created by a subliminal word. It enters the visual parts of the cortex, and travels through the visual areas of the ventral face of the brain. If the conditions are right, a subliminal word can even access higher levels of processing, including semantic levels. This is something that was highly controversial in psychology, but is now very clear from brain imaging: A subliminal message can travel all the way to the level of the meaning of the word. Your brain can take a pattern of shapes on the retina, and successively turn it into a set of letters, recognize it as a word, and access a certain meaning—all of that without any form of consciousness.

Next comes the obvious question: Where is there more activity when you are conscious of the word? If we do this experiment with fMRI, what we see is that two major differences occur. You first see an amplification of activation in the early areas: the very same areas begin to activate much more, as much as tenfold in, for instance, this area that we have been studying and which looks at the spelling of words: the visual word form area.

The second aspect is that several other distant areas of the brain activate. These include areas in the so-called prefrontal cortex, which is in the front of the brain here. In particular, we see activation in the inferior frontal region, as well as in the inferior parietal sectors of the brain. What we find also is that these areas begin to correlate with each other—they co-activate in a coordinated manner. I am for the moment just giving you the facts: Amplification and access to distant areas are some of the signatures of consciousness.

Now, if we look at a time-lapse picture, obtained with a technique such as electro-encephalography, which can resolve the temporal unfolding of brain activity, then we see something else which is very important: The difference between a nonconscious and a conscious

Stanislas Dehaene

percept occurs quite late in processing. Let's call time zero the point at which the word first appears on the screen, and let's follow this activation from that point. What we see is that, under the best of conditions, it can take as along as 270 to 300 milliseconds before we see any difference between conscious and unconscious processing.

For one-fourth of a second, which is extraordinarily long for the brain, you can have identical activations, whether you are conscious or not. During this quarter of a second, the brain is not inactive and we can observe a number of instances of lexical access, semantic access, and other processes (and subliminal processing can even continue after this point). But at about 270 milliseconds, 300 milliseconds in our experiments, we begin to see a huge divergence between conscious states and nonconscious states. If we only record using electrodes placed on the scalp, to measure what are called event-related potentials, we see a very broad wave called the P3 or P300. It's actually very easy to measure, and indeed one of the claims that my colleagues and I make is that access to consciousness is perhaps not that difficult to measure, after all.

The P3 wave is typically seen in conditions where subjects are conscious of the stimulus. You can get very small P3 waves under subliminal conditions, but there seems to be a very clear nonlinear divergence between conscious and nonconscious conditions at this point in time. When we manipulate consciousness using minimal contrasts, we find that subliminal stimuli can create a small and quickly decaying P3 wave, whereas a very big and nonlinear increase in activation, leading to a large event-related potential, can be seen when the same stimuli cross the threshold and become conscious.

At the same time as we see this large wave, which peaks at around 400, 500 milliseconds, we also see two other signatures of consciousness. First, our electrodes detect a high level of oscillatory activity in the brain, in the high-gamma band (50–100 hertz). Second, as the brain begins to oscillate at these high frequencies, we also begin to see massive synchrony across distant regions. What that means is that initially, prior to conscious ignition, processing is essentially modular, with several simultaneous activations occurring independently and in parallel. However, at the point where we begin to see conscious access, our records show a synchronization of many areas that begin to work together.

A Global Neuronal Workspace

I just gave you the bare facts: the basic signatures of consciousness that we have found and that many other people have also seen. I would now like to say a few words about what we think that these observations mean. We are on slightly more dangerous ground here, and I am sorry to say, a little bit fuzzier ground, because we cannot claim to have a full theory of conscious access. But we do begin to have an idea.

This idea is relatively simple, and it is not far from the one that Daniel Dennett proposed when he said that consciousness is "fame in the brain." What I propose is that "consciousness is global information in the brain"—information which is shared across different brain areas. I am putting it very strongly, as "consciousness is," because I literally think that's all there is. What we mean by being conscious of a certain piece of information is that it has reached a level of processing in the brain where it can be shared.

Because it is sharable, your Broca's area (or the part of it involved in selecting the words that you are going to speak) is being informed about the identity of what you are seeing, and you become able to name what you are seeing. At the same time, your hippocampus is perhaps informed about what you have just seen, so you can store this representation in memory. Your parietal areas also become informed of what you have seen, so they can orient attention, or decide that this is not something you want to attend to . . . and so on and so forth. The

Stanislas Dehaene

criterion of information sharing relates to the feeling that we have that, whenever a piece of information is conscious, we can do a very broad array of things with it. It is available.

Now, for such global sharing to occur, at the brain level, special brain architecture is needed. In line with Bernard Baars, who was working from a psychological standpoint and called it a "global workspace," Jean-Pierre Changeux and I termed it the global neuronal workspace. If you look at the associative brain areas, including dorsal parietal and prefrontal cortex, anterior temporal cortex, anterior cingulate, and a number of other sites, what you find is that these areas are tightly intertwined with long-distance connections, not just within a hemisphere, but also across the two hemispheres through what is called the corpus callosum. Given the existence of this dense network of long-distance connections, linking so many regions, here is our very simple idea: These distant connections are involved in propagating messages from one area to the next, and at this very high level where areas are strongly interconnected, the density of exchanges imposes a convergence to a single mental object out of what are initially multiple dispersed representations. So this is where the synchronization comes about.

Synchronization is probably a signal for agreement between different brain areas. The areas begin to agree with each other. They converge onto a single mental object. In this picture, each area has its own code. Broca's area has an articulatory record and slightly more anterior to it there is a word code. In the posterior temporal regions, we have an acoustic code, a phonological code, or an orthographic code. The idea is that when you become aware of a word, these codes begin to be synchronized together, and converge to a single integrated mental content.

According to this picture, consciousness is not accomplished by one area alone. There would be no sense in trying to pinpoint consciousness in a single brain area, or in computing the intersection of all the images that exist in the literature on consciousness, in order to find the area for consciousness. Consciousness is a state that involves long-distance synchrony between many regions. And during this state, it's not just higher association areas that are activated, because these areas also amplify, in a top-down manner, the lower brain areas that received the sensory message in the first place.

I hope that I have managed to create a mental picture for you of how the brain achieves a conscious state. We simulated this process on the computer. It is now possible to simulate neural networks that are realistic in the sense that they include actual properties of transmembrane channels, which are put together into fairly realistic spiking neurons, which in turn are put together into fairly realistic columns of neurons in the cortex, and also include a part of the thalamus, a subcortical nucleus which is connected one-to-one to the cortex. Once we put these elements together into a neural architecture with long-distance connections, we found that it could reproduce many of the properties that I described to you, the signatures of consciousness that we observed in our empirical observations.

So when we stimulate this type of network, at the periphery, we actually see activation climbing up the cortical hierarchy, and if there is enough reverberation and if there are enough top-down connections, then we see a nonlinear transition toward an ignited state of elevated activity. It's of course very simple to understand: In any dynamic system that is self-connected and amplifies its own inputs, there is a nonlinear threshold. As a result, there will be either a dying out of the activation (and I claim that this state of activity corresponds to subliminal processing), or there will be self-amplification and a nonlinear transition to this high-up state where the incoming information becomes stable for a much longer period of time. That, I think, corresponds to what we have seen as the late period in our recordings where the activation is amplified and becomes synchronized across the whole brain. In essence, very simple

Stanislas Dehaene

simulations generate virtually all of the signatures of consciousness that we've seen before.

What Is Consciousness Good For?

This type of model may help answer a question that was difficult to address before—namely, what is consciousness useful for? It is a very important question, because it relates to the evolution of consciousness. If this theory is right, then we have a number of answers to what consciousness actually does. It is unnecessary for computations of a Bayesian statistical nature, such as the extraction of the best interpretation of an image. It seems that the visual brain does that in a massively parallel manner, and is able to compute an optimal Bayesian interpretation of the incoming image, thus coming up with what is essentially the posterior distribution of all the possible interpretations of the incoming image. This operation seems to occur completely nonconsciously in the brain. What the conscious brain seems to be doing is amplify and select one of the interpretations, which is relevant to the current goals of the organism.

In several experiments, we have contrasted directly what you can do subliminally and what you can only do consciously. Our results suggest that one very important difference is the time duration over which you can hold on to information. If information is subliminal, it enters the system, creates a temporary activation, but quickly dies out. It does so in the space of about one second, a little bit more perhaps depending on the experiments, but it dies out very fast anyway. This finding also provides an answer for people who think that subliminal images can be used in advertising, which is of course a gigantic myth. It's not that subliminal images don't have any impact, but their effect, in the very vast majority of experiments, is very short-lived. When you are conscious of information, however, you can hold on to it essentially for as long as you wish. It is now in your working memory, and is now meta-stable. The claim is that conscious information is reverberating in your brain, and this reverberating state includes a self-stabilizing loop that keeps the information stable over a long duration. Think of repeating a telephone number. If you stop attending to it, you lose it. But as long as you attend to it, you can keep it in mind.

Our model proposes that this is really one of the main functions of consciousness: to provide an internal space where you can perform thought experiments, as it were, in an isolated way, detached from the external world. You can select a stimulus that comes from the outside world, and then lock it into this internal global workspace. You may stop other inputs from getting in, and play with this mental representation in your mind for as long as you wish.

In fact, what we need is a sort of gate mechanism that decides which stimulus may enter, and which stimuli are to be blocked, because they are not relevant to current thoughts. There may be additional complications in this architecture, but you get the idea: a network that begins to regulate itself, only occasionally letting inputs enter.

Another important feature that I have briefly mentioned already is the all-or-none property. You either make it into the conscious workspace or you don't. This is a system that discretizes the inputs. It creates a digital representation out of what is initially just a probability distribution. We have some evidence that, in experiments where we present a stimulus that is just at threshold, subjects end up either seeing it perfectly and completely with all the information available to consciousness—or they end up not having seen anything at all. There doesn't seem to be any intermediate state, at least in the experiments that we've been doing.

Having a system that discretizes may help solve one of the problems that John von Neumann considered as one of the major problems facing the brain. In his book *The Computer and the Brain*, von Neumann discusses the fact that the brain, just like any other analog

Stanislas Dehaene

machine, is faced with the fact that whenever it does a series of computations, it loses precision very quickly, eventually reaching a totally inaccurate result in the end. Well, maybe consciousness is a system for digitizing information and holding on to it, so that precision isn't lost in the course of successive computations.

One last point: Because the conscious workspace is a system for sharing information broadly, it can help develop chains of processes. A speculative idea is that once information is available in this global workspace, it can be piped to any other process. What was the output of one process can become the input of another, thus allowing for the execution of long and purely mental algorithms—a human Turing machine.

In the course of evolution, sharing information across the brain was probably a major problem, because each area had a specialized goal. I think that a device such as this global workspace was needed in order to circulate information in this flexible manner. It is extremely characteristic of the human mind that whatever result we come up with, in whatever domain, we can use it in other domains. It has a lot to do, of course, with the symbolic ability of the human mind. We can apply our symbols to virtually any domain.

So if my claim is correct, whenever we do serial processing, one step at a time, passing information from one operation to the next in a completely flexible manner, we must be relying on this conscious workspace system. This hypothesis implies that there is an identity between slow serial processing and conscious processing—something that was noted, for instance, by the cognitive psychologist Michael Posner already many years ago.

We have some evidence in favor of this conclusion. Let me tell you about a small experiment we did. We tried to see what people could do subliminally when we forced them to respond. Imagine that I flash a digit at you, and this digit is subliminal because I have masked it. Now suppose that I ask you, "Is it larger or smaller than five?" I give you two buttons, one for larger and one for smaller, and I force you to respond. When I run this experiment, although you claim that you have not seen anything, and you have to force yourself to respond, you find that you do much better than chance. You are typically around 60 percent correct, while pure chance would be 50 percent. So this is subliminal processing. Some information gets through, but not enough to trigger a global state of consciousness.

Now, we can change the task to see what kind of tasks can be accomplished without consciousness. Suppose we ask you to name a digit that you have not seen by uttering a response as fast as you can. Again, you do better than chance, which is quite remarkable: your lips articulate a word which, about 40 percent of the time, is the correct number, where chance, if there are four choices, would be 25 percent.

However, if we now give you a task that involves two serial processing steps, you cannot do it. If I ask you to give me the number plus two, you can do it—but if I ask you to compute the initial number plus two, and then decide if the result of that +2 operation is larger or smaller than five, you cannot do it. It's a strange result, because the initial experiments show that you possess a lot of information about this subliminal digit. If you just named it, you would have enough information to do so correctly, much better than chance alone would predict. However, when you are engaged in a chain of processing, where you have to compute $x+2$, and then decide if the outcome is larger or smaller than five, there are two successive steps that make performance fall back down to chance. Presumably, this is because you haven't had access to the workspace system that allows you to execute this kind of serial mental rule.

With a group of colleagues, we are working on a research project called "the human Turing machine." Our goal is to clarify the nature of the specific system, in the mind and in the brain, which allows us to perform such a series of operations, piping information from one

Stanislas Dehaene

stage to the next. Presumably the conscious global workspace lies at the heart of this system.

The Ultimate Test

Let me close by mentioning the ultimate test. As I mentioned at the beginning, if we have a theory of consciousness, we should be able to apply it to brain-lesioned patients with communication and consciousness disorders. Some patients are in coma, but in other cases, the situation is much more complicated. There is a neurological state called the vegetative state, in which a patient's vigilance—namely, the capacity to waken—can be preserved. In those patients there is a normal sleep-wake cycle, and yet, there doesn't appear to be any consciousness in the sense that the person doesn't seem to be able to process information and react normally to external stimulation and verbal commands.

There are even intermediate situations of so-called minimal consciousness, where a patient can, on some occasions only, for some specific requests, provide a response to a verbal command, suggesting that there could be partially preserved consciousness. On other occasions, however, the patient does not react, just as in the vegetative state, so that in the end we don't really know whether or not his occasional reactions constitute sufficient evidence of the patient's consciousness or not. Finally, as you all know, there is the famous "locked in" syndrome. It is very different in the sense that the patient is fully conscious, but this condition can appear somewhat similar in that the patient may not be able to express that he is conscious. Indeed, the patient may remain in a totally noncommunicative state for a long time, and it may be quite hard for others to discern that he is, in fact, fully aware of his surroundings.

With Lionel Naccache, we tried to design a test of the signatures of consciousness, based on our previous observations, that can indicate, very simply, in only a few minutes, based on observed brain waves alone, whether or not there is a conscious state. We opted for auditory stimuli, because this input form, unlike the visual modality, allows you to simply stimulate the patient without having to worry about whether he is looking at an image or not. Furthermore, we decided to use a situation that is called the mismatch response. Briefly, this means that the brain can react to novelty, either in a nonconscious or in a conscious way. We think that the two are very easy to discriminate, thanks to the presence of a late wave of activation which signals conscious-level processing.

Let me just give you a very basic idea about the test. We stimulate the patient with five tones. The first four tones are identical, but the fifth can be different. So you hear something like dit-dit-dit-dit-tat. When you do this, a very banal observation, dating back twenty-five years, is that the brain reacts to the different tone at the end. That reaction, which is called mismatch negativity, is completely automatic. You get it even in coma, in sleep, or when you do not attend to the stimulus. It's a nonconscious response.

Following it, however, there is also, typically, a later brain response called the P3. This is exactly the large-scale global response that we found in our previous experiments, that must be specifically associated with consciousness.

How to separate the two kinds of brain response is not always easy. Typically, they unfold in time, one after the other, and in patients, it is not always easy to separate them when they are distorted by a brain lesion. But here is a simple way to generate the P3 alone. Suppose the subject becomes accustomed to hearing four "dits" followed by a single "tat." What we see is that there still is a novelty response at the end, but it is now one that is expected. Because you repeat four "dits" followed by one "tat," over and over again, the subject can develop a conscious expectation that the fifth item will be different. What does the brain then do? Well, it still generates an early automatic and

Stanislas Dehaene

nonconscious novelty response—but it then cancels its late response, the P3 wave, because the repeated stimulus is no longer attracting attention or consciousness.

Now, after adaptation, the trick consists in presenting five identical tones: dit-dit-dit-dit-dit. This banal sequence becomes the novel situation now—the rare and unexpected one. Our claim is that only a conscious brain can take into account the context of the preceding series of tones, and can understand that five identical tones are something novel and unexpected.

In a nutshell, this outcome is exactly what we find in our experiments. We find a large P3-like response to five identical tones, in normal subjects after adaptation to a distinct sequence, but only in those subjects who attend to the stimulus and appear to be conscious of it. We tested the impact of a distracting task on normal subjects, and the P3 response behaves exactly as we expected. If you distract the subject, you lose this response. When the subject attends, you keep it. When the subject can report the rule governing the sequence, you see a P3. When they cannot report it, you don't see it.

Finally, Lionel Naccache, at the Salpêtrière Hospital, and Tristan Bekinschtein, now in Cambridge, moved on to applying these findings to patients. What they have shown is that the test would appear to behave exactly as we would expect. The P3 response is absent in coma patients. It is also gone in most vegetative state patients—but it remains present in most minimally conscious patients. It is always present in locked-in patients and in any other conscious subject.

The presence of this response in a few vegetative state patients makes one wonder if the person is really in a vegetative state or not. For the time being this is an open question—but it would appear that the few patients who show this response are the ones who are recovering fast and are, in fact, so close to recovery at the time of the test that you might wonder whether they were conscious during that test or not.

In summary, we have great hopes that our version of the mismatch test is going to be a very useful and simple test of consciousness. You can do it at the bedside—after ten minutes of EEG acquisition, you already have enough data to detect this type of response.

The Future of Neuroimaging Research: Decoding Consciousness

This is where we stand today. We have the beginnings of a theory, but you can see that it isn't yet a completely formal theory. We do have a few good experimental observations, but again there is much more we must do. Let me now conclude by mentioning what I and many other colleagues think about the future of brain imaging in this area.

My feeling is that the future lies in being able to decode brain representations, not just detect them. Essentially all that I have told you today concerns the mere detection of processing that has gone all the way to a conscious level. The next step, which has already been achieved in several laboratories, is decoding which representation is retained in the subject's mind. We need to know the content of a conscious representation, not just whether there is a conscious representation. This is no longer science fiction. Just two weeks ago, Evelyn Eger, who is a postdoc in Andreas Kleinschmidt's team in my laboratory, showed that you can take functional MRI images from the human brain and, by just looking at the pattern of activation in the parietal cortex, which relates to number processing, you can decode the number the person has in mind. If you take 200 voxels in this area, and look at which of them are active and which are inactive, you can construct a machine-learning device that decodes which number is being held in memory.

I should probably say quite explicitly that this use of the verb "decode" is an exaggeration. All you can do at the moment is achieve a better-than-chance inference about the memorized number. It does

Stanislas Dehaene

not mean that we are reading the subject's mind on every trial. It merely means that, whereas chance would be 50 percent for deciding between two digits, we manage to achieve 60 or 70 percent. That's not so bad, actually. It's a significant finding, which means that there is a cerebral code for number, and that we now understand a little bit more about that code.

But let me return again to patients. What I am thinking is that, in the future, with this type of decoding tool, we will move on to a new wave of research, where the goal will be explicitly to decode the contents of patients' minds, and maybe allow them to express themselves with the help of a brain-computer interface. Indeed, if we can decode content, there is no reason why we could not project it on the computer and use this device as a form of communication, even if the patient can no longer speak.

Again, some of this research has already started, in my lab as well as in several other places in the world. With Bertrand Thirion, we have looked at the occipital areas of the brain where there is a retinotopic map of incoming visual images. We have shown that you can start from the pattern of activation on this retinotopic map and use it to decode the image a person is viewing. You can even infer, to some extent, what mental image he has in his mind's eye, even when he is actually facing a blank screen. Mental images are a reality—they translate into a physical pattern of activation on these maps that you can begin to decode. Other researchers, such as Jack Gallant, in the U.S., have now pursued this research program to a much greater extent.

I believe that the future of neuroimaging lies in decoding a sequence of mental states, trying to see what William James called the stream of consciousness. We should not just decode a single image, but a succession of images. If we could literally see this stream, it would become even easier, without even stimulating the subject, to see that he is conscious because such a stream is present in his brain.

Don't mistake me, though. There is a clear difference between what we have been able to do—discover signatures of consciousness—and what I hope that we will be able to do in the future—decode conscious states. The latter remains very speculative. I just hope that we will eventually get there. However, it is already a fact that we can experiment on consciousness and obtain a lot of information about the kind of brain state that underlies it.

17

How Can Educated People Continue to Be Radical Environmentalists?

David Lykken (1928–2006)

Behavioral Geneticist, University of Minnesota; Author, Happiness: The Nature and Nurture of Joy and Contentment

How is it that some scientists, psychologists like Leon Kamin, biologists like Steven Rose, even the odd geneticist like Richard Lewontin, or the odd paleontologist like Stephen Gould, continue to believe with John Locke that the infant human mind is a tabula rasa. How can they suppose that baby brains are as alike as new Macintosh computers fresh from the factory; indeed, even more alike because the computers at least have operating systems and various ROMs already installed? How can anyone imagine that, sometime in the Pleistocene, evolution mysteriously stopped, but just for one subsystem of one mammalian genus, the nervous system of the genus *Homo*?

Without postulating that we possess ancestral inclinations, slowly acquired over many millennia, how could one explain why children tend to shy away from snakes and spiders but not from guns or electric sockets, which are much more dangerous? When the Minnesota Twins won the World Series in 1987 and again in 1991, when "our boys" had defeated those invaders from the National League, why did nearly 4 million Minnesotans, most of whom had never seen a game, proudly think that something wonderful had happened? When the Gulf War ended and "our boys" had killed a lot of Iraqis so the sultan of Kuwait could return from the Riviera to rebuild his palaces, the entire U.S. Congress stood, some on their seats or desks, to cheer

President Bush for his accomplishment. Those senators and representatives were not play-acting to impress their constituents; they really felt proud (but why?).

Romantic love, which anthropologists once thought had been invented by French poets in the Middle Ages, is now known to have characterized virtually every traditional society of which we have records. The other great apes do not experience infatuation because they do not need to pair bond. The baby chimps cling to their mother's fur and she can provide for their care and sustenance without any help from the unknown father. But when our ancestors began producing those big-headed, altricial babies that needed several years of constant carrying and oversight, more than the mothers could manage on their own, some sort of attachment had to be invented to persuade the fathers to help out. It turns out that, over all known societies that permit divorce, the modal length of marriage for those couples who eventually split is just four years; the fast-setting superglue of romantic infatuation lasts just long enough for Junior to be sturdy on his feet.

Identical twins, whose tastes are remarkably similar in all other respects, are about as likely to be charmed by their cotwin's romantic choice as by some passing stranger of the same age and gender. The spouses of identical twins, infatuated with Twin A at the time they meet Twin B, are no more inclined to "fall for" Twin B, the clone of their beloved, than for the boy (or girl) across the street. Natural selection had millions of years in which to fashion pair bonding in eagles and wolves, but it was a hurry-up job for the early hominids. My guess is that the mechanism used in our case was similar to that which produces imprinting in ducks and geese.

In their London debate, "The Two Steves" (Pinker and Rose) alluded briefly to why human parents love their babies. Pinker, a sensible evolutionary psychologist, thinks it is probably because those ancestral parents who were not somehow motivated to nurture their offspring were unlikely to have grandchildren and thus to become ancestors. I was never clear about what Rose thinks. But a more interesting question is why do Americans spend billions annually on dogs and cats and other pets? Assuming Pinker is correct, as assuredly he is, would natural selection continue fiddling with the machinery until parents felt nurturant about their own genetic offspring only? For some seals and sea birds that operate giant collective nurseries, where the young may wander off from their mothers, it appears that both mothers and offspring have evolved olfactory methods of identifying one another. But for most mammals, including the featherless bipeds, the danger of a parent "wasting" effort nurturing an unrelated baby was low enough so that a more precise targeting of maternal affection was unnecessary. The selection pressure favoring the more discriminating mothers was not great enough to produce a species change. Natural selection is parsimonious. It continues just long enough to fashion the ROM or module required to accomplish the necessary result in the environment of evolutionary adaptation.

A recent news report tells of a lost dog that had been fitted with a radio collar and was finally located in the den of a mother bear. Each time the dog started to emerge in response to his master's call, the bear gently drew him back again to his new home. My wife and I, like millions of others of our species, are more like the bears than we are like seals in this respect. In most jurisdictions, a person who kills a neighbor's dog or cat is treated by the law like someone who destroyed the neighbor's lawn mower. If lawmakers understood evolutionary psychology (or human pet owners) better, the offense would be treated much more seriously. My bull terrier is to me much more like my adopted child, if I had one, than like my lawn mower.

Another example of the parsimony of natural selection is our human xenophobia. We tend to distrust, fear, and dislike other humans who seem different than ourselves. This was adaptive in ancestral times, when a stranger stepped out from behind a tree because that stranger might kill you, if you were a male, or to rape you or carry you off, if you were a female. A more selective mechanism would require both strangeness and threatening action to trigger our fear/dislike response. But just "stranger" turned out to be enough. Can this be why modern humans, both New Yorkers and the natives of Papua New Guinea, tend to paint and dress themselves in ways that immediately identify their group membership? And can this be one of the reasons why our contemporary Lockians want to believe in the tabula rasa mythology? "Let us not suppose that xenophobia is natural because then how could we hope to accomplish racial, religious, and social tolerance?" My three sons, white, non-theistic, Aryan types, are happily married to a Catholic, a Jew, and an African American, and have produced my ten beloved grandchildren. Reporting some early results of his celebrated study of twins who were separated in infancy and reared apart, my colleague Tom Bouchard pointed out that: "The genes sing a prehistoric song that must sometimes be resisted but which should never be ignored." Our xenophobia *can* be resisted, as my sons' example attests, but it should not be ignored, or we shall never be able to figure out what to do about Bosnia.

Another ancestral trait that we should not ignore is male sexual jealousy. One human sex difference that even Gloria Steinem cannot deny is that a woman knows that the baby she delivered is hers while her spouse cannot be certain it is his. (DNA data indicate, in fact, that about 10 percent of human children could not have been produced by the mother's husband; for bluebirds, by comparison, the figure is about 20 percent.) The CEO of Natural Selection is aware of these facts and has endowed the males of our species with a suite of compensatory tendencies. David Buss has shown that, over many cultures, women are more disturbed by evidence that their mate has an affectionate relationship with another woman (an attachment that might lead him to invest his resources in her and her children), while men are much more concerned to learn that their mate has had sex with someone else. I once did some marriage counseling with a young "hippy" couple who were having problems. Their deeply held principles included opposition to the Vietnam War, support of environmental protections, legalization of psychoactive drugs, and free love. Their problem was that the young man was always grouchy and resentful. The solution to their problem was to accept the fact that most men cannot help feeling grouchy and resentful when their mate persists in having sex with other men. "Oh, baby, I'm so sorry! I didn't think you cared!"

Another curious fact is that even some evolutionary psychologists, including Steve Pinker's mentors, John Tooby and Leda Cosmides, believe that the genetic differences between people, the very differences on which, during ancestral times, natural selection worked to make us what we are today, no longer exist. "Yes, we all come equipped with species-specific behavioral proclivities. Our infant brains are not just general-purpose computers waiting to be programmed by experience but, rather, they have modules that are preprogrammed to give us a head start at being human. But they are all alike at birth, except perhaps for bits of noisy artifact." Are these folks just being politic, just claiming only the minimum they need to pursue their own agenda while leaving the behavior geneticists to contend with the main armies of political correctness?

The denial of genetically based psychological differences is the kind of sophisticated error normally accessible only to persons having PhD degrees. Even the be-doctored tend to give up radical environmentalism once they have a second child. In our twenty-five years of twin research at Minnesota, monozygotic twins, who share all their genes, have been found to be twice (or more than twice) as similar as dizygotic twins, who share on average half their polymorphic genes, on nearly every trait that we can measure reliably. The few exceptions include birth weight, years of education, romantic choice, and a few interests such as blood sports, gambling, and religious orientation. (Variation in general religiosity, on the other hand, is strongly genetic.) Moreover, monozygotic (MZ) twins separated in infancy and reared apart are as similar on most psychological traits as are MZ twins reared together. Middle-aged MZ twins, whether reared together or apart, correlate in IQ more than .70, and this is so whether IQ is estimated from the nonverbal Raven Matrices test administered and scored by computer, or from a standard IQ test individually administered by different examiners in separate rooms. IQ is not all there is to "intelligence" but it is very important. If your child's IQ is less than about 115, she is almost certain never to get through medical or law school.

One of the personality inventories that we use has a Well-Being scale that measures current happiness. Like most psychological traits (even IQ), happiness varies from time to time due to the slings and arrows. When we measure Well-Being in adult twins twice, ten years apart, the within-twin cross-time or retest correlation is only .55 (.02). But for MZ twins, the between-twin cross-time correlation (Twin A now versus Twin B then, etc.) is virtually the same, .54 (.03), suggesting that most of the happiness "set-point" or stable component is genetically determined. In contrast, the between-twin cross-time correlation for dizygotic (DZ) twins is only .05 (.07).

Happiness is one of the interesting traits that I call "emergenic." Although they have strong genetic roots, hence the strong MZ correlations, the negligible similarity of DZ twins indicates that these traits do not tend to run in families. Metrical traits that do run in families, traits like stature, reflect the additive combination of polygenic effects (the lengths of the head, neck, torso, upper and lower leg add up to body height). Emergenic traits seem to involve configural rather than additive combinations of the polygenic effects, so that small gene changes can produce large changes in the trait. Because each parent contributes just half of her or his genes to each child,

　　　　　　　　　　　　　　　　　　　　　　　David Lykken

and because siblings share on average just half of their polymorphic genes, first-degree relatives are unlikely to share all of the genes involved in an emergenic configuration.

Facial beauty seems to be an emergenic trait, as is the distinctive quality of the singing or speaking voice. MZ twins can usually fool even family members by impersonating their cotwins on the telephone; DZ twins very seldom can do this. Music majors at my university, including those specializing in voice, commonly have musical parents, but the voice majors seldom have parents who sing. The racing ability of the legendary stallion Secretariat seems to have been emergenic. Mated with only the most promising mares, he produced more than 400 foals; only one of them (Risen Star) was a winner, and even he could not have run with Dad.

Were it not for ideological prejudice, any rational person looking at the evidence would agree that human aptitudes, personality traits, many interests and personal idiosyncrasies, even some social attitudes, owe from 30 to 70 percent of their variation across people to the genetic differences between people. The ideological barrier seems to involve the conviction that accepting these facts means accepting biological determinism, Social Darwinism, racism, and other evils. I myself fall prey to this mistake from time to time. In the paper reporting our happiness data, for example, noting that the happiness set-point is largely genetic, while the events that move us temporarily above or below our set-points are largely fortuitous, I wrote: "perhaps trying to be happier is like trying to be taller." To make up for this error, I have had to write a book (*Happiness: The Nature and Nurture of Joy and Contentment*) explaining why trying to be happier is both feasible and fun.

The actual mechanism by which the genes affect the mind is still what Pinker (and Noam Chomsky) would call a mystery rather than a mere problem. We do not have a clue about how the brain module that permits humans but not chimps to acquire language is actually fashioned by the genetic enzyme factory. In the comparatively simple brain of a chicken there is a gizmo that produces an alarm reaction when the silhouette of a flying hawk passes overhead, but not when it is passed backward so that it looks more like a flying chicken. We cannot locate that gizmo or describe its construction. We do not know which genes in the chicken DNA mutated eons ago to bring about this adaptive response, and we clearly have no idea at all as to how these genes manage to fabricate this gizmo in every modern chicken's brain. Yet the existence of the gizmo cannot be doubted.

In the case of most human psychological traits, however, an important part of the mechanism is less mysterious. We know that, to an important extent, the genes affect the human mind indirectly by influencing the kinds of experiences we have, the way in which other people react to us, and especially by influencing the kinds of environments we seek out and the ways in which we react to our experiences. For genetic reasons, some babies are fretful and unresponsive while others tend to smile and coo. These different behaviors elicit different parenting responses. A genetically venturesome toddler climbs on things, falls off, explores, knocks things over, and has physical and social experiences that his more sedentary sibling seldom has. A naturally bright, inquisitive youngster notices and thinks about things, reads more, asks more questions, and elicits better answers than does a child whose mental processes are slower and less intrinsically rewarding. A little boy who is at the low end of the normal distribution of genetic fearfulness is less easily intimidated by the punishment on which both parents and peers tend to rely in trying to modify that boy's behavior. Many parents of such children give up the battle and the child remains unsocialized, a kind of psychopath. More skillful and persistent parents emphasize reward instead of punishment, work to instill pride rather than guilt. A fearless child left to himself is likely to become a leader of the gang, delinquent, then criminal, but with skillful parenting that same boy can grow to be the kind of man

David Lykken

we like to have around when danger threatens. I think that the hero and the psychopath are twigs on the same genetic branch.

Thus, genetic effects on human psychology are often distal in the causal chain while the proximal causes are environmental, just as those reactionary Lockians have always claimed. A better formula than Nature versus Nurture would be Nature *via* Nurture. But, distal or not, the genetic influences are strong and most of us develop along a path determined mainly by our personal genetic steersmen. It is often possible to intervene but it is seldom easy. A genetically timid child, for example, can be desensitized by carefully calibrated exposures to increasingly stressful situations. Meanwhile, a genetically venturesome child, a boy like General Chuck Yeager, for example, is doing the same thing much faster on his own, climbing higher, taking greater risks, learning to fly, becoming a fighter pilot, then an ace, then a test pilot, and finally breaking the sound barrier.

In her book, *The Nurture Assumption*, Judy Harris argues that parents' contribution to what will be their children's adult personality, interests, and attitudes is substantially completed when sperm meets egg. The experiences that will interact with the genome to determine that child's adult future occur mainly outside the home, with the peer group. One reason young adults feel slightly uncomfortable going home at Thanksgiving, Harris suggests, is that as children they learned one way of behaving at home and another way outside with their peers, and it was the latter suite of habits and values that developed into their adult personas. Back home for a visit, they find themselves again wearing the parent-approved personality, a disguise they had discarded as children whenever they left the house and which they had thought was gone forever when they reached adulthood. This radical doctrine (to which I cannot do justice here, of course) dumps much of developmental psychology into the recycle bin, and it is bound to dismay all parents except those whose children haven't turned out very well.

Yet Harris's arguments are so cogent and compelling that I, for one, have been forced to reassess. She says, in effect, that if we plotted the success of children's socialization on the Y-axis, and the skill of the parents on the X-axis, then the function relating the two variables would be a horizontal line. Harris has convinced me that the curve really is flat (that parents really are fungible) in the broad middle range between, say, the 10 percent and the 90 percent points on the X-axis distribution of parental competence. (Harris points out that this 80 percent approval rating is better than Clinton's!) But I think (although I cannot prove) that the curve rises on the far right, that there are some super-parents who really do make a lasting difference, the parents who succeed in socializing the really difficult children, for example. And I am confident that the bottom 10 percent, the immature, abusive, unsocialized, or simply incompetent parents (which includes a large proportion of the rising tide of impoverished and overburdened single mothers) are responsible for the epidemic of crime and other social pathology that has been accelerating in this country since the 1960s.

In spite of small declines each year since 1993, the rate of violent crime in the U.S. is presently 300 percent higher than it was in 1960. The recent dip is due largely to the fact that there are now 1.3 million Americans in state or federal prisons, compared to about 180,000 in 1965. Because the average inmate will privately admit to some twelve crimes committed in the year prior to his last arrest, imprisoning an extra million men is bound to yield a small but significant decrease in the crime rate. But it is an expensive and an inadequate solution. The place to fight crime is in the cradle. My own proposal would be parental licensure along the lines suggested by the child psychiatrist Jack Westman, in his 1994 book with that title.

Once again, evolutionary psychology and behavior genetics can provide guidance. Traditional societies, in which children are reared much as our ancestors were, experience very little intramural crime.

The few outlaws in those communities tend to be people whose innate temperaments made them extraordinarily difficult to socialize, people we would now call psychopaths. We were designed by natural selection to be able to develop a conscience, feelings of empathy and altruism, to become responsible and to carry our share of the load in the group effort for survival. Like our language instinct, these socialization proclivities require to be elicited, shaped, and reinforced beginning in early childhood. In the extended-family milieu of our hunter-gatherer ancestors, with the help of numerous adults and the older children, we can suppose that this process was usually successful. When a modern young couple, inexperienced and untrained, attempt this most demanding of human responsibilities on their own, we can expect the failure rate to be higher. When a single mother, often immature and poorly socialized herself and usually in straightened circumstances, takes on this responsibility, the failure rate is very high indeed. In the U.S., more than two-thirds of incarcerated delinquents, teenage mothers, high school dropouts, teenage runaways, and juvenile murderers were reared without fathers.

Evolutionary psychology does not tell us that crime is biologically determined but, rather, the opposite. Behavior geneticists have never located any "crime genes," although they have identified heritable traits of temperament that make some children hard to socialize. Regarding this greatest social problem of our time, these two lines of research dictate a message, not of fatalism, but of hope. Those 1.3 million men now languishing in American prisons began as innocent babes, some of them difficult, most of them average, almost all of whom could have been fashioned into taxpaying citizens, friends, and neighbors, had they been luckier in the circumstances of their growing up.

18

Moral Psychology and the Misunderstanding of Religion

Jonathan Haidt

Psychologist, University of Virginia; Author, The Happiness Hypothesis

I study morality from every angle I can find. Morality is one of those basic aspects of humanity, like sexuality and eating, that can't fit into one or two academic fields. I think morality is unique, however, in having a kind of spell that disguises it. We all care about morality so passionately that it's hard to look straight at it. We all look at the world through some kind of moral lens, and because most of the academic community uses the same lens, we validate each other's visions and distortions. I think this problem is particularly acute in some of the new scientific writing about religion.

When I started graduate school at Penn in 1987, it seemed that developmental psychology owned the rights to morality within psychology. Everyone was either using or critiquing Lawrence Kohlberg's ideas, as well as his general method of interviewing kids about dilemmas (such as: Should Heinz steal a drug to save his wife's life?). Everyone was studying how children's understanding of moral concepts changed with experience. But in the 1990s two books were published that I believe triggered an explosion of cross-disciplinary scientific interest in morality, out of which has come a new synthesis—very much along the lines that E. O. Wilson predicted in 1975.

The first was Antonio Damasio's *Descartes' Error,* in 1994, which showed a very broad audience that morality could be studied using the then new technology of fMRI, and also that morality, and rationality itself, were crucially dependent on the proper functioning of emotional circuits in the prefrontal cortex. The second was Frans de Waal's *Good Natured*, published just two years later, which showed an equally broad audience that the building blocks of human morality are found in other apes and are products of natural selection in the highly social primate lineage. These two books came out just as John Bargh was showing social psychologists that automatic and unconscious processes can and probably do cause the majority of our behaviors, even morally loaded actions (like rudeness or altruism) that we thought we were controlling consciously.

Furthermore, Damasio and Bargh both found, as Michael Gazzaniga had years before, that people couldn't stop themselves from making up post hoc explanations for whatever it was they had just done for unconscious reasons. Combine these developments and suddenly Kohlbergian moral psychology seemed to be studying the wagging tail, rather than the dog. If the building blocks of morality were shaped by natural selection long before language arose, and if those evolved structures work largely by giving us feelings that shape our behavior automatically, then why should we be focusing on the verbal reasons that people give to explain their judgments in hypothetical moral dilemmas?

In my dissertation and my other early studies, I told people short stories in which a person does something disgusting or disrespectful that was perfectly harmless (for example, a family cooks and eats its dog, after the dog was killed by a car). I was trying to pit the emotion of disgust against reasoning about harm and individual rights.

I found that disgust won in nearly all groups I studied (in Brazil, India, and the United States), except for groups of politically liberal college students, particularly Americans, who overrode their disgust and said that people have a right to do whatever they want, as long as they don't hurt anyone else.

These findings suggested that emotion played a bigger role than the cognitive developmentalists had given it. These findings also suggested that there were important cultural differences, and that academic researchers may have inappropriately focused on reasoning about harm and rights because we primarily study people like ourselves—college students, and also children in private schools near our universities, whose morality is not representative of the United States, let alone the world.

So in the 1990s I was thinking about the role of emotion in moral judgment, I was reading Damasio, De Waal, and Bargh, and I was getting very excited by the synergy and consilience across disciplines. I wrote a review article called "The Emotional Dog and Its Rational Tail," which was published in 2001, a month after Josh Greene's enormously influential *Science* article. Greene used fMRI to show that emotional responses in the brain, not abstract principles of philosophy, explain why people think various forms of the "trolley problem" (in which you have to choose between killing one person or letting five die) are morally different.

Obviously I'm biased in terms of what I notice, but it seems to me that the zeitgeist in moral psychology has changed since 2001. Most people who study morality now read and write about emotions, the brain, chimpanzees, and evolution, as well as reasoning. This is exactly what E. O. Wilson predicted in *Sociobiology*: that the old approaches to morality, including Kohlberg's, would be swept away or merged into a new approach that focused on the emotive centers of the brain as biological adaptations. Wilson even said that these emotive centers give us moral intuitions, which the moral philosophers then justify while pretending that they are intuiting truths that are independent of the contingencies of our evolved minds.

And now, thirty years later, Josh Greene has a paper in press where he uses neuroscientific evidence to reinterpret Kantian deontological philosophy as a sophisticated post hoc justification of our gut feelings about rights and respect for other individuals. I think E. O. Wilson deserves more credit than he gets for seeing into the real nature of morality and for predicting the future of moral psychology so uncannily. He's in my pantheon, along with David Hume and Charles Darwin. All three were visionaries who urged us to focus on the moral emotions and their social utility.

I recently summarized this new synthesis in moral psychology with four principles:

1. *Intuitive primacy but not dictatorship.* This is the idea, going back to Wilhelm Wundt and channeled through Robert Zajonc and John Bargh, that the mind is driven by constant flashes of affect in response to everything we see and hear.

Our brains, like other animal brains, are constantly trying to fine-tune and speed up the central decision of all action: approach or avoid. You can't understand the river of fMRI studies on neuroeconomics and decision making without embracing this principle. We have affectively valenced intuitive reactions to almost everything, particularly to morally relevant stimuli such as gossip or the evening news. Reasoning by its very nature is slow, playing out in seconds.

Studies of everyday reasoning show that we usually use reason to search for evidence to support our initial judgment, which was made in milliseconds. But I do agree with Josh Greene that sometimes we can use controlled processes such as reasoning to override our initial intuitions. I just think this happens rarely, maybe in 1 or 2 percent of the hundreds of judgments we make each week. And I do agree with Marc Hauser that these moral intuitions require a lot of computation, which he is unpacking.

Hauser and I mostly disagree on a definitional question: whether this means that "cognition" precedes "emotion." I try never to contrast those terms, because it's all cognition. I think the crucial contrast is between two kinds of cognition: intuitions (which are fast and usually affectively laden) and reasoning (which is slow, cool, and less motivating).

2. *Moral thinking is for social doing.* This is a play on William James's pragmatist dictum that thinking is for doing, updated by newer work on Machiavellian intelligence. The basic idea is that we did not evolve language and reasoning because they helped us to find truth; we evolved these skills because they were useful to their bearers, and among their greatest benefits were reputation management and manipulation.

Just look at your stream of consciousness when you are thinking about a politician you dislike, or when you have just had a minor disagreement with your spouse. It's like you're preparing for a court appearance. Your reasoning abilities are pressed into service generating arguments to defend your side and attack the other. We are certainly able to reason dispassionately when we have no gut feeling about a case, and no stake in its outcome, but with moral disagreements that's rarely the case. As David Hume said long ago, reason is the servant of the passions.

3. *Morality binds and builds.* This is the idea stated most forcefully by Emile Durkheim, that morality is a set of constraints that binds people together into an emergent collective entity.

Durkheim focused on the benefits that accrue to individuals from being tied in and restrained by a moral order. In his book *Suicide* he alerted us to the ways that freedom and wealth almost inevitably foster anomie, the dangerous state where norms are unclear and people feel that they can do whatever they want.

Durkheim didn't talk much about conflict between groups, but Darwin thought that such conflicts may have spurred the evolution of human morality. Virtues that bind people to other members of the tribe and encourage self-sacrifice would lead virtuous tribes to vanquish more selfish ones, which would make these traits more prevalent.

Of course, this simple analysis falls prey to the free-rider problem that George Williams and Richard Dawkins wrote so persuasively about. But I think the terms of this debate over group selection have changed radically in the last ten years, as culture and religion have become central to discussions of the evolution of morality.

I'll say more about group selection in a moment. For now I just want to make the point that humans *do* form tight, cooperative groups that pursue collective ends and punish cheaters and slackers, and they do this most strongly when in conflict with other groups. Morality is what makes all of that possible.

4. *Morality is about more than harm and fairness.* In moral psychology and moral philosophy, morality is almost always about how people treat each other. Here's an influential definition from the Berkeley psychologist Elliot Turiel: Morality refers to "prescriptive judgments of justice, rights, and welfare pertaining to how people ought to relate to each other."

Kohlberg thought that all of morality, including concerns about the welfare of others, could be derived from the psychology of justice. Carol Gilligan convinced the field that an ethic of "care" had a separate developmental trajectory, and was not derived from concerns about justice.

OK, so there are two psychological systems, one about fairness/justice and one about care and protection of the vulnerable. And if you look at the many books on the evolution of morality, most of them focus exclusively on those two systems, with long discussions of Robert Trivers's reciprocal altruism (to explain fairness) and of kin altruism and/or attachment theory to explain why we don't like to see suffering and often care for people who are not our children.

But if you try to apply this two-foundation morality to the rest of the world, you either fail or you become Procrustes. Most traditional societies care about a lot more than harm/care and fairness/justice. Why do so many societies care deeply and morally about menstruation, food taboos, sexuality, and respect for elders and the Gods? You can't just dismiss this stuff as social convention. If you want to describe human morality, rather than the morality of educated Western academics, you've got to include the Durkheimian view that morality is in large part about binding people together.

From a review of the anthropological and evolutionary literatures, Craig Joseph (at Northwestern University) and I concluded that there were three best candidates for being additional psychological foundations of morality, beyond harm/care and fairness/justice. These three we label as *ingroup/loyalty* (which may have evolved from the long history of cross-group or subgroup competition, related to what Joe Henrich calls "coalitional psychology"); *authority/respect* (which may have evolved from the long history of primate hierarchy, modified by cultural limitations on power and bullying, as documented by Christopher Boehm); and *purity/sanctity*, which may be a much more recent system, growing out of the uniquely human emotion of disgust, which seems to give people feelings that some ways of living and acting are higher, more noble, and less carnal than others.

Joseph and I think of these foundational systems as expressions of what Dan Sperber calls "learning modules"—they are evolved modular systems that generate, during enculturation, large numbers of more specific modules that help children recognize, quickly and automatically, examples of culturally emphasized virtues and vices. For example, we academics have extremely fine-tuned receptors for sexism (related to fairness) but not sacrilege (related to purity).

Virtues are socially constructed and socially learned, but these processes are highly prepared and constrained by the evolved mind. We call these three additional foundations the *binding* foundations, because the virtues, practices, and institutions they generate function to bind people together into hierarchically organized interdependent social groups that try to regulate the daily lives and personal habits of their members. We contrast these to the two *individualizing* foundations (harm/care and fairness/reciprocity), which generate virtues and practices that protect individuals from each other and allow them to live in harmony as autonomous agents who can focus on their own goals.

My UVA colleagues Jesse Graham, Brian Nosek, and I have collected data from about 7,000 people so far on a survey designed to measure people's endorsement of these five foundations. In every sample we've looked at, in the United States and in other Western countries, we find that people who self-identify as liberals endorse moral values and statements related to the two individualizing foundations primarily, whereas self-described conservatives endorse values and statements related to all five foundations. It seems that the moral domain encompasses more for conservatives—it's not just about Gilligan's care and Kohlberg's justice. It's also about Durkheim's issues of loyalty to the group, respect for authority, and sacredness.

I hope you'll accept that as a purely descriptive statement. You can still reject the three binding foundations normatively—that is, you can still insist that ingroup, authority, and purity refer to ancient and dangerous psychological systems that underlie fascism, racism, and homophobia, and you can still claim that liberals are right to reject those foundations and build their moral systems using primarily the harm/care and fairness/reciprocity foundations.

But just go with me for a moment that there is this difference, descriptively, between the moral worlds of secular liberals on the one hand and religious conservatives on the other. There are, of course, many other groups, such as the religious left and the libertarian right, but I think it's fair to say that the major players in the new religion wars are secular liberals criticizing religious conservatives. Because the conflict is a moral conflict, we should be able to apply the four principles of the new synthesis in moral psychology.

In what follows I will take it for granted that religion is a part of the natural world that is appropriately studied by the methods of science. Whether or not God exists (and as an atheist I personally doubt it), religiosity is an enormously important fact about our species.

Jonathan Haidt

There must be some combination of evolutionary, developmental, neuropsychological, and anthropological theories that can explain why human religious practices take the various forms that they do, many of which are so similar across cultures and eras. I will also take it for granted that religious fundamentalists, and most of those who argue for the existence of God, illustrate the first three principles of moral psychology (intuitive primacy, post hoc reasoning guided by utility, and a strong sense of belonging to a group bound together by shared moral commitments).

But because the New Atheists talk so much about the virtues of science and our shared commitment to reason and evidence, I think it's appropriate to hold them to a higher standard than their opponents. Do these New Atheist books model the scientific mind at its best? Or do they reveal normal human beings acting on the basis of their normal moral psychology?

1. *Intuitive primacy but not dictatorship.* It's clear that Richard Dawkins (in *The God Delusion*) and Sam Harris (in *Letter to a Christian Nation*) have strong feelings about religion in general and religious fundamentalists in particular. Given the hate mail they receive, I don't blame them. The passions of Dawkins and Harris don't mean that they are wrong, or that they can't be trusted. One can certainly do good scholarship on slavery while hating slavery.

But the presence of passions should alert us that the authors, being human, are likely to have great difficulty searching for and then fairly evaluating evidence that opposes their intuitive feelings about religion. We can turn to Dawkins and Harris to make the case for the prosecution, which they do brilliantly, but if we readers are to judge religion we will have to find a defense attorney. Or at least we'll have to let the accused speak.

2. *Moral thinking is for social doing.* This is where the scientific mind is supposed to depart from the lay mind. The normal person (once animated by emotion) engages in moral reasoning to find ammuni-

tion, not truth; the normal person attacks the motives and character of her opponents when it will be advantageous to do so. The scientist, in contrast, respects empirical evidence as the ultimate authority and avoids ad hominem arguments. The metaphor for science is a voyage of discovery, not a war. Yet when I read the New Atheist books, I see few new shores. Instead I see battlefields strewn with the corpses of straw men. To name three:

a. The New Atheists treat religions as sets of beliefs about the world, many of which are demonstrably false. Yet anthropologists and sociologists who study religion stress the role of ritual and community much more than of factual beliefs about the creation of the world or life after death.

b. The New Atheists assume that believers, particularly fundamentalists, take their sacred texts literally. Yet ethnographies of fundamentalist communities (such as James Ault's *Spirit and Flesh*) show that even when people claim to be biblical literalists, they are in fact quite flexible, drawing on the Bible selectively—or ignoring it—to justify humane and often quite modern responses to complex social situations.

c. The New Atheists all review recent research on religion and conclude that it is an evolutionary by-product, not an adaptation. They compare religious sentiments to moths flying into candle flames, ants whose brains have been hijacked for a parasite's benefit, and cold viruses that are universal in human societies. This denial of adaptation is helpful for their argument that religion is bad for people, even when people think otherwise.

I quite agree with these authors' praise of the work of Pascal Boyer and Scott Atran, who have shown how belief in supernatural entities may indeed be an accidental output of cognitive systems that otherwise do a good job of identifying objects and agents. Yet even if belief in gods was initially a by-product, as long as such beliefs had consequences for behavior then it seems likely that natural selection operated upon phenotypic variation and favored the success of individuals and groups that found ways (genetic or cultural or both) to use these gods to their advantage—for example, as commitment devices that enhanced cooperation, trust, and mutual aid.

3. *Morality binds and builds.* Dawkins is explicit that his goal is to start a movement, to raise consciousness, and to arm atheists with the arguments they'll need to do battle with believers. The view that "we" are virtuous and our opponents are evil is a crucial step in uniting people behind a cause, and there is plenty of that in the New Atheist books. A second crucial step is to identify traitors in our midst and punish or humiliate them. There is some of that too in these books—atheists who defend the utility of religion or who argue for disengagement or détente between science and religion are compared to Chamberlain and his appeasement of Hitler.

To my mind an irony of Dawkins's position is that he reveals a kind of religious orthodoxy in his absolute rejection of group selection. David Sloan Wilson has supplemented Durkheim's view of religion (as being primarily about group cohesion) with evolutionary analyses to propose that religion was the conduit that pulled humans through a "major transition" in evolutionary history.

Dawkins, along with George Williams and most critics of group selection, acknowledge that natural selection works on groups as well as on individuals, and that group selection is possible in principle. But Dawkins relies on Williams's argument that selection pressures at the individual level are, in practice, always stronger than those at the group level: Free riders will always undercut Darwin's suggestion that morality evolved because virtuous groups outcompeted selfish groups.

Wilson, however, in *Darwin's Cathedral*, makes the case that culture in general and religion in particular change the variables in Williams's analysis. Religions and their associated practices greatly increase the costs of defection (through punishment and ostracism), increase the contributions of individuals to group efforts (through cultural and emotional mechanisms that increase trust), and sharpen the boundaries—biological and cultural—between groups. Throw in recent discoveries that genetic evolution can work much faster than previously supposed, and the widely respected work of Pete Richerson and Rob Boyd on cultural group selection, and suddenly the old consensus against group selection is outdated.

It's time to examine the question anew. Yet Dawkins has referred to group selection in interviews as a "heresy," and in *The God Delusion* he dismisses it without giving a reason. In chapter 5 he states the standard Williams free-rider objection, notes the argument that religion is a way around the Williams objection, concedes that Darwin believed in group selection, and then moves on. Dismissing a credible position without reasons, and calling it a heresy (even if tongue in cheek), are hallmarks of standard moral thinking, not scientific thinking.

4. *Morality is about more than harm and fairness.* In *Letter to a Christian Nation*, Sam Harris gives us a standard liberal definition of morality: "Questions of morality are questions about happiness and suffering. . . . To the degree that our actions can affect the experience of other creatures positively or negatively, questions of morality apply." He then goes on to show that the Bible and the Koran, taken literally, are immoral books because they're not primarily about happiness and suffering, and in many places they advocate harming people.

Reading Harris is like watching professional wrestling or the Harlem Globetrotters. It's great fun, with lots of acrobatics, but it must not be mistaken for an actual contest. If we want to stage a fair fight between religious and secular moralities, we can't eliminate one by definition before the match begins. So here's my definition of morality, which gives each side a chance to make its case:

Moral systems are interlocking sets of values, practices, institutions, and evolved psychological mechanisms that work together to suppress or regulate selfishness and make social life possible.

In my research I have found that there are two common ways that cultures suppress and regulate selfishness, two visions of what society is and how it ought to work. I'll call them the *contractual* approach and the *beehive* approach.

The contractual approach takes the individual as the fundamental unit of value. The fundamental problem of social life is that individuals often hurt each other, and so we create implicit social contracts and explicit laws to foster a fair, free, and safe society in which individuals can pursue their interests and develop themselves and their relationships as they choose.

Morality is about happiness and suffering (as Harris says, and as John Stuart Mill said before him), and so contractualists are endlessly trying to fine-tune laws, reinvent institutions, and extend new rights as circumstances change in order to maximize happiness and minimize suffering. To build a contractual morality, all you need are the two individualizing foundations: harm/care and fairness/reciprocity. The other three foundations, and any religion that builds on them, run afoul of the prime directive: Let people make their own choices, as long as they harm nobody else.

The beehive approach, in contrast, takes the group and its territory as fundamental sources of value. Individual bees are born and die by the thousands, but the hive lives for a long time, and each individual has a role to play in fostering its success. The two fundamental problems of social life are attacks from outside and subversion from within. Either one can lead to the death of the hive, so all must pull together, do their duty, and be willing to make sacrifices for the group. Bees don't have to learn how to behave in this way but human children do, and this is why cultural conservatives are so heavily focused on what happens in schools, families, and the media.

Conservatives generally have a more pessimistic view of human nature than do liberals. They are more likely to believe that if you stand back and give kids space to grow as they please, they'll grow into shallow, self-centered, undisciplined pleasure seekers. Cultural conservatives work hard to cultivate moral virtues based on the three binding foundations: ingroup/loyalty, authority/respect, and purity/sanctity, as well as on the universally employed foundations of harm/care and fairness/reciprocity. The beehive ideal is not a world of maximum freedom, it is a world of order and tradition in which people are united by a shared moral code that is effectively enforced, which enables people to trust each other to play their interdependent roles. It is a world of very high social capital and low anomie.

It might seem obvious to you that contractual societies are good, modern, creative, and free, whereas beehive societies reek of feudalism, fascism, and patriarchy. And, as a secular liberal I agree that contractual societies such as those of Western Europe offer the best hope for living peacefully together in our increasingly diverse modern nations (although it remains to be seen if Europe can solve its current diversity problems).

I just want to make one point, however, that should give contractualists pause: Surveys have long shown that religious believers in the United States are happier, healthier, longer-lived, and more generous to charity and to each other than are secular people. Most of these effects have been documented in Europe, too. If you believe that morality is about happiness and suffering, then I think you are obligated to take a close look at the way religious people actually live and ask what they are doing right.

Don't dismiss religion on the basis of a superficial reading of the Bible and the newspaper. Might religious communities offer us in-

Jonathan Haidt

sights into human flourishing? Can they teach us lessons that would improve well-being even in a primarily contractualist society?

You can't use the New Atheists as your guide to these lessons. The New Atheists conduct biased reviews of the literature and conclude that there is no good evidence on any benefits except the health benefits of religion. Here is Daniel Dennett in *Breaking the Spell* on whether religion brings out the best in people:

> Perhaps a survey would show that as a group atheists and agnostics are more respectful of the law, more sensitive to the needs of others, or more ethical than religious people. *Certainly no reliable survey has yet been done that shows otherwise.* It might be that the best that can be said for religion is that it helps some people achieve the level of citizenship and morality typically found in brights. *If you find that conjecture offensive, you need to adjust your perspective.* (*Breaking the Spell*, p. 55)

I have italicized the two sections that show ordinary moral thinking rather than scientific thinking. The first is Dennett's claim not just that there is no evidence, but that there is *certainly* no evidence, when in fact surveys have shown for decades that religious practice is a strong predictor of charitable giving. Arthur Brooks recently analyzed these data (in *Who Really Cares*) and concluded that the enormous generosity of religious believers is not just recycled to religious charities.

Religious believers give more money than secular folk to secular charities, and to their neighbors. They give more of their time, too, and of their blood. Even if you excuse secular liberals from charity because they vote for government welfare programs, it is awfully hard to explain why secular liberals give so little blood. The bottom line, Brooks concludes, is that all forms of giving go together, and all are greatly increased by religious participation and slightly increased by conservative ideology (after controlling for religiosity).

These data are complex and perhaps they can be spun the other way, but at the moment it appears that Dennett is wrong in his reading of the literature. Atheists may have many other virtues, but on one of the least controversial and most objective measures of moral behavior—giving time, money, and blood to help strangers in need— religious people appear to be morally superior to secular folk.

My conclusion is *not* that secular liberal societies should be made more religious and conservative in a utilitarian bid to increase happiness, charity, longevity, and social capital. Too many valuable rights would be at risk, too many people would be excluded, and societies are so complex that it's impossible to do such social engineering and get only what you bargained for. My point is just that every longstanding ideology and way of life contains some wisdom, some insights into ways of suppressing selfishness, enhancing cooperation, and ultimately enhancing human flourishing.

But because of the four principles of moral psychology it is extremely difficult for people, even scientists, to find that wisdom once hostilities erupt. A militant form of atheism that claims the backing of science and encourages "brights" to take up arms may perhaps advance atheism. But it may also backfire, polluting the scientific study of religion with moralistic dogma and damaging the prestige of science in the process.

Jonathan Haidt

BOOKS BY JOHN BROCKMAN

CULTURE
ISBN 978-0-06-202313-1 (paperback)

THE MIND
ISBN 978-0-06-202584-5 (paperback)

IS THE INTERNET CHANGING THE WAY YOU THINK?
ISBN 978-0-06-202044-4 (paperback)

THIS WILL CHANGE EVERYTHING
ISBN 978-0-06-189967-6 (paperback)

WHAT HAVE YOU CHANGED YOUR MIND ABOUT?
ISBN 978-0-06-168654-2 (paperback)

WHAT ARE YOU OPTIMISTIC ABOUT?
ISBN 978-0-06-143693-2 (paperback)

WHAT IS YOUR DANGEROUS IDEA?
ISBN 978-0-06-121495-0 (paperback)

WHAT WE BELIEVE BUT CANNOT PROVE
ISBN 978-0-06-084181-2 (paperback)